Sanaa MAJID

Ultra-sensitive electrochemical sensors

Sanaa MAJID

Ultra-sensitive electrochemical sensors

Trace heavy metal detection and analysis

ScienciaScripts

Cover image: www.ingimage.com

This book is a translation from the original published under ISBN 978-620-6-70419-5.

Publisher:
Sciencia Scripts
is a trademark of
Dodo Books Indian Ocean Ltd. and OmniScriptum S.R.L publishing group

120 High Road, East Finchley, London, N2 9ED, United Kingdom
Str. Armeneasca 28/1, office 1, Chisinau MD-2012, Republic of Moldova, Europe
Printed at: see last page
ISBN: 978-620-8-30324-2

Contents

GENERAL INTRODUCTION

In recent years, legislation on water and food testing has become increasingly stringent. This is due not only to the constant concern to protect human health, but also to the rapid increase in the number of sources of contamination threatening the environment. The cumulative effect of these compounds must also be taken into account, and their long-term effects monitored.

As environmental awareness grows, so does the need for accurate, reliable measurements. As stricter limits are imposed to protect people, regular checks must be carried out to ensure safety.

To meet these legitimate demands, complex scientific research is carried out, and its effectiveness is closely linked to the quality of the analytical tools made available.

Faced with these challenges, laboratories need to design and develop increasingly effective methods for analysing trace substances. These analytical techniques must combine specificity and high sensitivity. The electrochemical approach, long neglected, has seen renewed interest over the last fifteen years with the advent of electrochemical sensors.

Electrochemistry offers attractive prospects for the miniaturisation and low-cost industrialisation of simple, accurate and robust sensors. However, in order to achieve levels of sensitivity comparable to those achieved by spectroscopic techniques, a great deal of research is currently under development.

Conventional ëlectrodes (Hg, Pt, Au, C) are subject to certain parasitic phënomënes limiting their field of application in both analysis and svntliese. The most important of these are, on the one hand, the dëgradation of the electrode surface consëcutive to precipitation or adsorption processes that are not dësirës and, on the other hand, the extreme slowness of certain ëlectrochemical transformations for which a high overvoltage must be applied. Furthermore, while a large number of ëlectroactive molecules are likely to be dëtectëed, the fact remains that many analytically interesting substances are generally little or not reactive at the level of conventional electrodes.

As a result, the concept and practical application of chemically modified electrodes has recently emerged to overcome these drawbacks. Studies into these chemically modified electrodes have multiplied and diversified and are still in full swing today, as several hundred studies attest.

Our laboratory is particularly interested in the development of new electrochemical methods, based on low-cost modified electrodes and designed to detect low levels of heavy metals. To this end, we have studied three different approaches for the analysis of heavy metals. The first approach consists of modifying the surface of a platinum electrode with an amino acid, L-tyrosine, and studying the response of the electrode to mercury. The second approach involves modifying the carbon paste electrode with an amine-containing monomer. Finally, the third approach is to use carbon screen-printed electrodes for the determination of lead and cadmium.

After a review of the literature in the first chapter on the characteristics of heavy metals and the various analytical and electrochemical methods of detecting them, we will give an overview of the various methods of modifying solid electrodes and carbon paste electrodes that exist in the literature.

In Chapter 2, we will review the theoretical and practical foundations on which the electrochemical techniques used throughout this work are based.

In Chapter 3, we will describe an amperometric method for the determination of traces of mercury. This determination will be based on the influence of the presence of mercury (II) ions on the oxidation current of the amino acid L-tyrosine at the surface of a platinum electrode.

In Chapter 4, we will insert the 1,8-DAN monomer into the carbon paste and study its polymerisation in different acidic media. We will then try out the application of this modified electrode as a chemical sensor capable of preconcentrating Pb(II) ions in an open circuit.

Finally, the development of a screen-printed electrode modified with a mercury acetate salt will be described in chapter 5. The application of this electrode to the simultaneous detection of lead and cadmium will be carried out on pharmaceutical samples.

PART I

THEORETICAL PART

Chapter I: Heavy metals and how they are detected

1- CHARACTERISTICS OF HEAVY METALS

I-1 Introduction

In дёпёrаl mëtaux lourds the ëlëments mëtaШques natural, mëtaux or in some cases mëtalloides caractërisës by a density ëlevëe, supëérieur a 5 grammes per cm^3 [1]. The term "heavy mëtal" is used in some older publications. Heavy metals are present in all environmental compartments, but generally in very small quantities. Heavy metals are said to be present "in trace amounts". They are also "traces" of the geological past and of human activity.

However, the term "heavy metals" is a common name that has no scientific basis or legal application. Most of these metals are toxic when present in concentrations or quantities above the admissible threshold.

The classification as heavy metals is often disputed because some toxic metals are not particularly "heavy" (zinc), while some toxic elements are not all metals (arsenic, for example) [2]. For these various reasons, most scientists prefer to refer to heavy metals as "metallic trace elements" or, by extension, "trace elements".

There are three main heavy metals: mercury, lead and cadmium. Early biochemists distinguished these three metals on the basis of their affinity with sulphur, which makes it possible to identify proteins that easily precipitate or give rise to salts (mercury salts, lead salts, etc.).

The three metals also share a number of physico-chemical characteristics:

- They are not destroyed. They are transported and change chemical form.
- They have high electrical conductivity, which explains their use in many industries.
- They are to some extent toxic to humans, causing neurological damage of varying severity [3]. All the others are useful in biological processes - some metals (trace elements) are even essential to life (iron, copper, nickel, chromium, etc.).

The three metals mentioned are toxic elements, and their chemical characteristics are as follows:

	Lead	Cadmium	Mercure
Atomic mass	270	112	200
Density	11.35 g/cm^3	8.6 g/cm^3	13.6 g/cm^3
Melting temperature	327°	320.9°	-38°
Boiling temperature	1740°	765°	357°
Chemical symbol	Pb	Cd	Hg
Original ore	Galene	Zinc slag	Cinnabar

I-2 Mercury

It is rare in the natural environment. However, traces of it can be found in rocks, sometimes in concentrations that justify mining. Mercury is notably extracted from cinnabar (mercury sulphate). Like lead, mercury has been used since ancient times. Its ability to combine with other metals was used to extract gold. Mercury has also been used for its biological properties, including its toxic properties (as biocides). It has ёle ийНзё in tanneries, agriculture, mëdicine to treat syphilis for example. Today mercury is ийНзё for its physico-chemical propriëtës. Mercury is extremely volatile, reacts with heat and is also an excellent conductor of electrical energy. It is used in the production of chlorine and some consumer or measurement products (batteries, thermometers, etc.). These uses are declining because of the pollution problems caused by this type of industry. The world produces around 3,000 tonnes of mercury a year.

I-2-1 Characteristics of mercury

Mercury is a metal with rare characteristics; it is the only metal that is liquid at room temperature (between -10° and 40°C). A characteristic of or is 11rë its chemical symbol Hg, from the Greek word Latinisë hydragyrm, liquid silver. When shaken, it divides into fine droplets. It is also the only metal with a boiling point below 655°C. This metal is extremely volatile. It combines easily with other modules, whether metals (amal games), inorganic modules (sulphur) or organic modules (carbon). It is also a toxic metal, its Ю\юlё coming from its extreme volatility (it can be easily breathed in), its relative solubility in water and fats (it can be easily transported in the body) and its ability to bind with other modules that it will modify or whose functions it will transform.

I-2-2 Forms of mercury

Physicochemically, mercury is a metal that changes form and properties easily. Highly volatile, it changes easily from a liquid to a gaseous state at room temperature. It also has a high expansion capacity, hence its use in both heavy and liquid thermometers or as a reference material for measuring pressure (barometers, tensiometers). In the presence of oxygen, mercury oxidises very rapidly.

It is easily converted from the mëtaШcic state (Hg°), liquid or gaseous, to the ionised state (Hg^{2+}). It is also a metal that associates easily with organic molecules, forming numerous mercurial dërivës. Mercury comes in two distinct families:

J Metallic or inorganic mercury, which itself takes three different forms:

- Elemental metallic mercury, in liquid form (note Hg°), is the classic mercury, in its best-known form, used in thermometers or for practical chemistry work (Polarography).
- Gaseous mercury (Hg°)
- Inorganic mercury, in ionic form (note Hg).$^{2+}$

J The other large family is organic mercury, when it combines with a molecule containing carbon, the basis of all living (or formerly living) elements. There are constant exchanges between these different forms, because mercury has a great capacity to transform, particularly under the effect of the acidity of the environment, and the presence of molecules or ligands that ensure these combinations (chlorine, sulphur).

I-2-3 Mercury toxicity

The toxicity of mercury varies according to its chemical form: - Mercury in liquid form (Hg°) is not very toxic because it is very little absorbed orally. Almost all (over 99%) of the mercury ingested leaves the body via the natural route (faeces, urine). A director of a poison control centre in Vienna personally underwent the experiment of swallowing 100 grams of metallic mercury: the mercury went into the stomach, then into the appendix. The level of mercury in the urine rose to 80 mg/litre after two months, then fell again until it was completely absorbed.

- The mercury mëtaШque in vapour form (Hg°) is no longer тдёгё (in the stomach) but inhaled, and consequently goes into the lungs and bloodstream [4]. The mercury is then transported to various parts of the body, including the brain, the target organ for mercury vapour poisoning. When the vapours originate from dental amalgam, some are swallowed, solubilised in saliva and absorbed by the stomach.

- Ionised mercury can enter the body orally (by inhalation) or through the skin [5]. This explains why the roots of teeth can be loaded with mercury after an amalgam filling [6]. These pathways are concentrated in the liver and kidneys.
- Organic mercury that has already been absorbed and assimilated by a living organism is found in its carbon tissues. It can be ingested again by another organism (e.g. mercury absorbed by fish and shellfish, concentrated in the digestive tract, which is then consumed by humans). This form is highly toxic [7,8]. A typical example is Minamata Bay in Japan [9]. Releases of methyl mercury were accumulated by fish eaten by the bay's inhabitants (in 1956: 549 victims, in 1965: 119 victims and a total of 1,200 deaths). Methyl mercury compounds are soluble in fat and can therefore cross cell membranes. Mercury was used by the Chisso factory as a catalyst in the production of acetaldehyde. In all, between 1932 and 1968, 81 tonnes of Hg were discharged into the bay.

Apart from direct discharges of methyl mercury (as in the case of Minamata), discharges of mercury in metallic form oxidise and then, in the aquatic environment under the action of bacteria, are transformed into methyl mercury, which is itself absorbed by plankton and then concentrates in fish. It is estimated that 5% of the mercury brought into the Mediterranean each year is found in fish. A phënomëne of biomagnification (successive concentrations) amplified the risk, since the concentrations in fish were 100,000 times higher than the concentration in seawater. Maximum mercury concentrations in shellfish and fish reached 179 mg/kg dry weight (WHO standard 2.5 mg/kg dry weight) and 23 mg/kg fresh weight (WHO standard 0.5 mg/kg fresh weight) respectively. Mercury concentrations in patients' hair can reach 705^g/kg (WHO standard 10^g/kg).

It is important to note that mercury can pass from the mother's blood to that of the ЬёЬё during pregnancy or breastfeeding [10]. Often irreversible effects can affect the child's nervous system [11]. Among other things, the child may suffer from cё^гeк paralysis, involuntary movements and impaired vision and hearing.

The phënomëne of biomagnification represents the main danger of mëthyl mercury, because from an apparently unpolished medium, the successive concentration can lead to concentrations a million times higher than the initial one, and therefore becomes very toxic.

I-3 Lead

I-3-1 Uses

Produced from a mineral called gakne, the use of lead is directly related to metallurgy. Lead was used extensively in the production of coins, pipes and crockery. During the first half of the 20th century^{eme} skcle, lead was used in industry, printing and paints. In the second half of the 20eme skcle, the dominant use was Hëo automotive fuels, with lead added to petrol as an antidëtonant. This use is now prolific.

I-3-2 The main toxic effects of lead

Lead poisoning refers to all the symptoms of lead intoxication. Lead colic is the best-known toxic effect of lead, but its main target organs are the nervous system, kidneys and blood. On the nervous system, lead is responsible for neurological damage. In the case of massive intoxication, the neurotoxic effect of lead can lead to convulsive encephalopathy

and even death. In the case of less severe intoxication, neurobehavioural disorders and intellectual deterioration have been observed.
In the bone marrow and blood, it blocks several enzymes needed to synthesise haemoglobin. These blood effects lead to a reduction in the number of red blood cells and to anaemia.
Administration of high doses of lead has induced kidney cancer in small rodents. However, there is no evidence of excess cancer mortality in populations exposed to lead. Acute intoxication^ is rare. Lead poisoning is usually due to chronic exposure.
In children, the risk of lead poisoning is higher, particularly between the ages of 1 and 3:
- Digestive absorption of lead derivatives is greater than in adults. With the same exposure, a child's body absorbs 50% of the lead ingested, whereas the proportion in adults is only 5 to 7%;
- the toxic effects, with equal impregnation, particularly on the developing central nervous system, are greater and more severe.

The central nervous system of children is particularly sensitive to the toxic action of lead. Acute convulsive encephalopathy generally occurs when blood lead levels are around 1000 gg/l. It has never been observed when blood lead levels are below 700 gg/l. In children whose blood lead levels are between 500 and 700 gg/l, less severe neurological disorders are often observed: reduced motor activity, irritability, sleep disorders, behavioural changes and stagnation in intellectual development. A drop in IQ of between 4 and 15 points!

I-4 Cadmium

I-4-1 Uses

It is a natural element, present in certain ores (particularly zinc) in the form of impurities. This metal was unknown until the 10th century, when its physico-chemical properties were brought to light and used in batteries in particular. Cadmium has been used extensively to protect steel against corrosion (cadmium plating) and as a stabiliser for plastics and pigments.
Cadmium is used in many alloys, as well as in the manufacture of batteries, cables, ball bearings, welding rods, fluorescents, dyes, medicines and pesticides. Cadmium is also present in certain fertilisers and is found in significant quantities in tobacco leaves.
The burning of coal and oil contributes to the accumulation of cadmium throughout the environment. Once dëposë, cadmium is absorbed by plants, some of which are destined for human consumption, such as Ыё or tegumes. Other contaminated plants serve as food for animals, which then concentrate the cadmium in their organs. Offal (liver, kidneys) is the edible part of the animal that poses the greatest risk to humans.

I-4-2 Health consequences

The various cadmium compounds have variable toxic effects [12] depending on how soluble they are and how easily they are absorbed by the body. Thus, cadmium chloride, which is soluble, appears to be more toxic than cadmium sulphide, which is very insoluble.
Short-term exposure to high concentrations of cadmium compounds in dusts or fumes is irritating to the cells of the respiratory and gastrointestinal systems [13].
The main causes of cadmium exposure are diet and smoking. Most cadmium тдёгё comes

from vëgëtaux with green foliage, lettuce, cabbage, ëpinach and to a lesser extent from cërëales. As cadmium accumulates mainly in the kidneys, this organ is considered to be a "target" organ.

People who eat a diet rich in offal (particularly certain types of game), and who also smoke, are likely to be affected by this metal.

Cadmium salts, which are not very volatile, are present in the air in the form of very fine solid particles (smoke or dust). During occupational exposure, these particles can be inhaled and are deposited mainly in the pulmonary alveoli. The size of these particles is therefore of great importance in determining their absorbability.

I-5 Heavy metal discharges into water

Industry has often favoured sites close to rivers for three reasons: for the transport of raw materials, for the supply of water to cool installations, and for the possibility of discharging industrial effluents. For decades, rivers have inherited industrial discharges and industrial wastewater, liquid waste resulting from the extraction or processing of raw materials and from all forms of production activity. Although some industrial establishments have their own wastewater treatment plants, the bulk of wastewater is discharged directly, sometimes referred to as "natural waste". For a long time, water - from rivers, canals and the sea - was the "outlet" for this waste.

Trace elements, present in particulate form in soils, appear as a result of erosion. Runoff from impermëaвles surfaces (soils and pavements) and anthropogenic sources are added to these natural sources linked to erosion.

This toxicity is reinforced by a concentration phënomëne known as bioaccumulation or biomagnification. Bioaccumulation is the process by which heavy metals are assimilated and concentrated in the body. The process takes place in two stages: bioaccumulation begins with the individual (soluble mercury is eliminated very little and is assimilated by the individual, animal, fish, etc.) and continues through transmission between individuals, by successive "stacks" (herbivore, fish-eater, etc.). Concentrations increase as you move up the trophic chain. This is known as biomagnification.

Water is obviously a particularly important element for pollutants in general, and heavy metals in particular, because water causes chemical reactions linked to acidity, alkalinity, temperature, oxygenation, etc. Aquatic environments are highly sensitive to trace elements due to the coexistence of two phenomena, bioaccumulation and biomagnification: trace elements concentrate as they are absorbed into the food chain (water plankton herbivorous fish carnivorous fish man).

Industry is responsible for almost all discharges of heavy metals into water. The need to reduce these discharges is no longer debated.

The analysis of heavy metal discharges into water is still under-researched and is much less closely monitored than nitrogen and phosphorus, for example.

In recent years, legislation has become increasingly stringent. This is due not only to the constant concern to protect human health, but also to the rapid increase in the number of sources of contamination threatening the environment. It is also necessary to monitor their long-term effect and to take account of their cumulative effect, given that heavy metals are micropollutants of a nature to cause nuisance even when they are discharged in very small quantities, their toxicity developing by bioaccumulation.

The French **NF** standards and the requirements of the World Health Organisation **WHO** relating to the characteristics of drinking water are grouped together in the table below:

Metal	NF	WHO
Hg	**1PPb**	**1PPb**
Pb	**10ppb**	**10ppb**
Cd	**5ppb**	**3ppb**

II- DETECTION OF HEAVY METALS

II-1 Introduction

Constant advances in materials technology and microelectronics are making a major contribution to the development of increasingly sophisticated instrumentation. Modern analytical equipment is integrated into an automated, computer-controlled system that allows exhaustive processing of experimental data. In the light of the considerable technological advances observed over the last ten years in spectral methods of analysis, such as atomic absorption spectroscopy and AAS, and in detection techniques coupled with separation systems such as high-performance liquid chromatography (HPLC), gas chromatography and capillary electrophoresis (CE) [14,17], it is interesting to consider the place occupied by electrochemical techniques.

II-2 Analytical methods for analysing heavy metals

Many analytical techniques are used to analyse heavy metals. Spectrophotomëtric techniques occupy a prominent place. The technique that has made it possible to achieve very low detection thresholds is atomic absorption spectrometry (AAS) [18-19].

Several spëciation or dëtermination studies of heavy metals have ël.ë rëalisëes, using cold vapour phase atomic absorption spectromëtrie (CVAAS) [20-21].

Plasma demission spectrometry (ICP-AES) is an analytical method that competes with atomic absorption spectrometry, as its detection limits are lower than those of SAA. This spectral method of analysis is based on the principle of emission: when atoms are subjected to energy conditions that allow them to be excited, spontaneous transitions from an excited ënergëtic level to a less excited ënergëtic level are accompanied by the emission of photons of frequency v and energy qk. The ionisation source is an argon plasma with induced current (T > 8000°C) maintained by ëenergy supplied in the form of a ë^й^^ field by a radiofrequency gënërateur. fkkment exc^ produces a spectrum of characteristic wavelengths whose luminous intensity is proportional to its concentration.

However, other methods are also applicable, such as flame emission spectrometry, molecular absorption spëctrophotometry and emission spectrography. These techniques do not always fulfil certain conditions such as multiplic^ of assays, simplic^ of the opëratory mode, precision, sensi bi l Hë and ra|жШë of exëcution.

Chromatographic techniques have also been used for the analysis of heavy metals. Cappon et al have used gas chromatography for the determination of inorganic and organic mercury in various biological media [22].

Several ëresearch teams have ийНзё gas chromatography with an electron capture dëtector (GC/ECD) for the analysis of methylmercury [23,24]. High-performance liquid chromatography (HPLC) has also been successfully used for the determination of heavy metals [25,26].

In the light of progress in the analytical field, spectral methods were coupled with chromatographic techniques, which led to an increase in sensitivity. For example, Ritsema's team [27] combined gas chromatography with atomic fluorescence spectrometry by hydride generation for the determination of mercury in seawater. High-performance liquid chromatography coupled with cold vapour atomic absorption spectrometry has been used by Munaf for the spëciation of mercury in waste water [28]. However, these techniques have the disadvënient of being very expensive techniques that lack flexibility and do not allow simultaneous analysis of the compounds. Hence the advantage of using electrochemical techniques.

II-3 Electrochemical analysis methods

II-3-1 Introduction

The diversity^ and sensitivity^ of the analyses currently required have forced the expërimentalist to use more precise and more efficient techniques, capable of providing qualitative and quantitative results in complex media, also allowing the investigation of several composës during a single opëration. The ëlectrochemical mëthods respond gënëralement to these cri teres while offering, among other advantages, excellent precision and easy and rapid handling.

In order to compete with the most sensitive analytical methods, electrochemical techniques have proved to be very sensitive and efficient, especially thanks to the development of electrochemical sensors, microelectrodes and modified electrodes... ***II-3-2 Electrochemical sensors for detecting heavy metals***

Electrochemical sensors are commonly prepared on either a metal or carbon base. As a general rule, the working electrode used must be stable over a very long period, have a good signal-to-noise ratio and be easy to handle and condition. In addition, the solutes to be analysed must develop rapid electrochemical reaction kinetics in a wide accessible potential range [29,30].

Metals such as platinum, gold, silver and steel have long been used because of their excellent electrical and mechanical properties. These electrodes are very sensitive to adsorption phenomena; nevertheless, by applying an electrochemical cleaning of the electrode surface, platinum and gold electrodes have been successfully used for the analysis of heavy metals [31-34].

At the same time, carbon is an increasingly popular material, due to its low cost, the variety of commercial forms available (graphite, carbon black, carbon fibre, etc.) and its flexibility in processing. Carbon-based materials are increasingly chemically inert and can be used over a wide range of potentials. The carbon-based electrodes commonly used in electrochemistry laboratories are vitreous carbon electrodes and carbon paste electrodes.

Carbon paste electrodes are relatively easy to produce, as they involve mixing carbon powder and a binder (paraffin oil, for example). These electrodes were introduced by R.N.Adams in 1958 [35]. The properties of this type of electrode dëpend on the purity of the composës ийНзёз to prepare the paste, the carbon/binder ratio [36,37], the size of the graphite particles [38], and the treatment and method of renewing the electrode surface [39,40].

The large-scale production of low-cost carbon-based electrochemical sensors has therefore been a priority in recent years. Most of these electrodes are single-use and are now produced using the serigraphy technique. Derived from microelectronics, the screen-printing of electrochemical sensors has developed rapidly since the early nineties [41]. Its simplicity, high production capacity and the possibility of producing sensors of any size or shape make the screen printing technique one of the most promising technologies [42]. These electrodes are obtained by depositing several alternating layers of carbon, silver and insulating inks on an insulating polyester substrate.

The use of these screen-printed electrodes offers a number of advantages, including ease of use, relatively low cost and the ability to take measurements in very small volumes. In fact, they meet all the simplicity and speed requirements of the medical, environmental and quality control sectors. These qualities make it possible to envisage portable systems, offering the possibility of carrying out measurements in the field [43].

The increasing use of carbon-based screen-printed electrodes is also due to the fact that they have enabled the development of many modified electrodes.

II-3-3 Modified electrodes

One of the major thrusts of new electroanalytical methods is the development of modified electrodes [44-51]. Although many applications can be envisaged using conventional ëlectrodes, their direct use in complex media comes up against surface problems (fouling by large molecules, surfactants, grease, etc....) and a lack of selectivity [52]. In order to resolve these problems, new strategies aimed at modifying the electrode surface were envisaged from the early 1980s [53,54]. The main aim of these strategies is to extend their field of application to the study of molecules that are not very reactive, or not at all, and to increase both the selectivity and the sensitivity of measurements [55-58], by giving them new characteristics by adding chemical or biological species to their surface (electroactive compounds, polymëres, biomolecules, etc.). There is a great deal of research in this field, as the possibilities for surface modification are so varied.

There are two main strategies for modifying electrodes:

a. Modification of the surface of solid electrodes chemically or electrochemically *(Tableaul)*

This strategy has been adopted by several authors for the determination of heavy metals.

- ***<u>Golden film</u>***

Viltchinskaia and colleagues were the first to modify the graphite electrode with a gold film in order to determine mercury using the LSASV technique [59]. Several other works on the determination of the same metal were carried out subsequently, by modifying the carbon paste electrode and graphite, and made it possible to achieve very low detection limits using ASV [60-61].

The gold film has also ële ийНзё by some authors for the determination of 1 arsenic by modifying different ëlectrodes such as glassy carbon and platinum electrode using LSASV [62.63]. The most sensitive method (LD= 0.02^g/l) was developed by Zakharova and co-workers who used the graphite electrode modified with a gold film for the determination of As(III) in drinking water and mineral water [64].

- ***<u>Mercury film</u>***

The mercury film has been widely used for the determination of several heavy metals. Solid electrodes, such as silver and platinum, have been used as substrates for the mercury film. However, the glassy carbon electrode remains the most suitable for this modification, due to its non-porous surface and ease of cleaning. Studies have shown that these electrodes can be used in a variety of media and allow the simultaneous determination of several trace metals. A study carried out by Petrov and colleagues enabled the simultaneous detection of Pb, Cd, Cu and Zn using SWASV in groundwater, with detection limits of 0.2, 0.01, 0.7, 2.1gg/l respectively [65]. The carbon paste electrode was also used by Svancara and co-workers as a substrate for the mercury film for the determination of Zn(II) in drinking water, deionised water and other samples using the DPASV technique [66]. Hart and colleagues have even modified the carbon screen electrode with a mercury film for the analysis of Cu^{2+} , Cd^{2+} and Pb^{2+} in urine using DPASV [67].

- ***Polymer***

One of the great advances of our time, in the field of chemistry, lies in the ability to synthesise new molecules for a specific function. This mastery of molecules has enabled the development of a scientific field of fundamental interest: "The modification of electrode surface properties".

Investigation in the field of modified ëlectrodes has been carried out along three main lines of approach:

The simplest modification is based on the exclusion of the various compounds analysed according to their size and the porosity of the polymers is involved [68-70]. Various studies have been carried out on cellulose acetate polymer to modify electrodes intended for the analysis of traces of heavy metals [71,72].

Another method of modification is based on the principle of ion exchange and involves the cationic or anionic nature of the polymer. The most commonly used polymers are perfluorosulphonate ionomers, such as Eastman-Kodak and Nafion. The latter polymer has been widely and successfully used. Thanks to its ability to exchange cations, various metals have been dosed [73-77]. Anion exchange polymers have also been used as electrode modifiers to analyse heavy metals. For example, Tosflex has been used for the analysis of mercury and copper [78,79] and poly (4- vinylpyridine) [80-82].

The third approach involves forming a polymer film within which a ligand capable of forming a complex with the metal to be analysed is grafted. The structure of the film therefore has a considerable effect on the sensitivity and selectivity of the sensor. Several studies have been published using different polymers. Polypyrrole has been used for electrode modification [83]. Song et al synthesised a film of polypyrrole on glassy carbon by electropolymerisation and used it for silver analysis, the detection limit was estimated to be around 0.2ppm. Copper and mercury were the main interferents during analysis [84]. Other polymers were also used, to which specific molecules were grafted in order to obtain chemically fixed active groups and ensure complexation. Zen et al. modified the glassy carbon electrode with Nafion, a chelating agent (dimethyl-glyoxime, 2,2-bipyridyl) and a mercury film; for the analysis of Cd, Pb and Cu in wastewater, drinking water, seawater and urine [85] . Polyaniline has also been used to modify glassy carbon eiectrode, and the polymer has been used as a matrix in which 2,2'-bipyridyl has been incorporated for lead analysis [86].

Polymers already containing functional groups allowing complex formation have also been used [87-89]. Palys et al studied the sensitivity of 1,8 diaminonaphthalene to heavy metals, showing that Cu^{2+} , Hg^{2+} ions can be complexed by the amine groups of the polymer [90]. In another study using DPV, trace amounts of Se(IV) were determined after modification of the gold electrode with a poly (3,3'-diaminobenzidine) film [91].

Tableaul: Modified solid electrodes for heavy metal analysis.

Taken from: **Kh. Z. Brainina et al " Stripping voltammetry in environmental and food analysis "** [92]

Electrode	Modification procedure	Analyst	Detection limit/ concentration determined (ug'L	Method used	ref
CV	Golden film	As(III)	0.5	LS ASV	[62]
Graphite	Golden film	As(III)	0.02	LS ASV	[64]
Graphite	Golden film	Hg(II)	0.1	LS ASV	[61]
Pt	Golden film	As(III), As(III+V)	0.5	LS ASV	[63]
CV	Mercury film	Pb(II) Cd(II)	0.2 0.0005	SQW ASV	[68]
CV	Mercury film	Pb(II) Cd(II) Cu(II) Zn(II)	0.2 0.01 0.7 2.1	SQW ASV	[65]
Ag	Mercury film	Pb(II) Cu(II) Cd(II)	10	LS ASV	[93]
Ag	Mercury film	Sn(IV) Pb(II)	0.5 0.1	LS ASV	[94]
Pt	Mercury film	Cu(II) Pb(II) Cd(II)	49.2 21.8 6.2	DP ASV	[95]
CV	Tosflex	Hg(II)	0.004	DP ASV	[80]
CV	Tosflex	Cu(II)	600	LS ASV	[81]
CV	Cationic polypyrrole derivative	Hg(II)	0.02	DP ASV	[87]
CV	8,9,17,18-dibenzo-1,7-dioxa-10,13,16-triazacyclooctadecane	Au(III)	16.4	SQW ASV	[88]
CV	Kryptofix-222	Hg(II)	<0.2	SQW ASV	[89]
CV	Nafion and dimethyl-dithiocarbamic acid(I) or 18-crown-6(II)	Pb(II) Cu(II) Hg(II)	5.10^{-7} -5.10^- 5mol/l (I) 2.10^{-6} - 4.10^{-5} mol/l (II)	DP ASV	[78]
Graphite	Nafion + chelating agent (1-(2-pyridylazo)-2-naphthol)	Pb(II) Cu(II)	1-2	SQW ASV	[79]
CV	Poly(4-vinylpyridine) + mercury film	Pb(II)	0.3	SQW ASV	[82]
CV	Nafion + chelating agent (dimethylglyoxime, 2,2-	Cu(II) Pb(II) Cd(II)	4 2	SQW ASV	[85]

	bipyridyl) + mercury film				
CV	Gold film/ Poly(4-vinyl-pyridine)	Hg(II)	0.1	SQW ASV	[83]
Gold	Poly(3,3'-diaminobenzidine)	Se(IV)	0.78	DP ASV	[91]
CV	Polypyrrole	Ag(I)	2ɪɪM	CV	[74]
CV	Polyaniline +2 ,2-bipyridyl	Pb(II)		ASV	[86]
Gold	1,8-diaininonaphthalene	Ag(I) Cu(II) Hg(II)		CV	[87]

b. Direct insertion of the modifying agent

This is the most widely used strategy for modifying carbon paste electrodes and screen-printed electrodes. It consists of adding the modifier to the mass of carbon paste for carbon paste electrodes or to the mass of carbon ink before screen printing for screen printed electrodes.

In general, the main reason for modifying an electrode is to qualitatively obtain a new sensor with desired and often predefined properties. For this purpose, carbon paste represents one of the most convenient materials for the preparation of modified electrodes [96,97]. In contrast to solid electrodes, the preparation of chemically modified carbon paste is very straightforward and is achieved by means of several alternative procëdures. The modifier can be dissolved directly from a mechanical fagon with the carbon powder [98-100] or be mëlangë with the paste during its homogënëisation [100102]. As it is also possible to dip the graphite particles in a modifier solution, and use the modified paste after ëvaporation of the solvent [103,104].

In 1984, Wang et al ël ë were the first to modify CPE with a cation-exchange ëresin and to study the analytical performance of this ëlectrode for Cu ion analysis^{2+} [105]. Subsequently several ëtudes, using the same mëthode have ël ë published.

Klacher has made a classification of four possible functions of modifiers used within the dough: [96]

- A prëfërential accumulation of dësirëe species via the modifying agent (e.g. preconcentration in redissolution analysis)
- Mëdiation reactions at the leetrode surface using immobilised molecules or their fragments.
- Catalysis of reactions impossible on the bare electrode and made possible by the modifying agent (catalytic electrochemical responses)
- Modification of the surface characteristics of the carbon paste electrode.

When you combine the functions of the modifying agent with the aforementioned flexibility and ease of processing of carbon paste, it is not surprising that over the last decade an increasing number of different substances have been used to modify carbon paste electrodes.

Bulk modification of screen-printed electrodes involves adding the modifying agent directly to the ink before screen printing and is based on the same principles as the

modification of carbon paste electrodes.
The modifying agents recently ий^ёз can be simple chemical agents [106-110], ë ion exchangers [111-113], zëolites [114], humic substances [115-117], silicas [118] and even living organisms [119,120].
Tables 2 and 3 summarise the various modifications and their applications in the analysis of heavy metals.

Table 2: Modifications to the carbon paste electrode for metal analysis heavy

Amendment procedure	Analyst	Detection limit / concentration dëterminëe (pg/l)	Mëthode used	Rëf
Ion exchanger (HYPHAN)	Cu(II) Pb(II) Hg(II)	15 23 10	DP ASV	[111]
Silver (I) Hexacyano-ferrate (III)	Fe (III)	120	DP ASV	[106]
Humic acid йхё by l^thylenediamine	To (III)	10	DP ASV	[115]
Phenylfluorone	Sb (III)	1	DP ASV	[107]
Humic acid amides	Hg (I, II)	10	LS ASV	[116]
Zn-DDC	Hg (II)	0.16	DP ASV	[121]
Clay	Hg (II)	10	DP ASV	[114]
Glyoxal bis (2-hydroxyanil)	Hg(II) Ag (I)	0.2 0.01	DP ASV	[122]
8-hydroxyquinoline	Ti(I)	1	DP ASV	[108]
Dithizone	Pb(II)	16.7	DP ASV	[123]
N-benzol-N',N' -di-i-butyl-thiourea	Ag (I)	10	LS ASV	[109]
Humic acid	Pb(II) Cu(II) Hg(II)	1 0.5 1.6	DP ASV	[117]
Dowex 50W-8X cation exchange rësine	Cu (II)	6.4	LS ASV	[112]
Chelite P	Cu (II)	1.9	DP ASV	[124]
Mercury film	Zn (II)	8.5	DP ASV	[66]

Table 3: Modifications to sërigraphiëes carbon ëlectrodes for the analysis of mëtaux lourds

Amendment procedure	Analyst	Detection limit / Dëterminëe concentration (LI g/L)	Mëthode used	ref
Mercury film	Cu(II),Pb(II), Cd(II)	0.05(Cd) 0.03(Pb)	DP ASV	[67]
Alizarin powder (5%) / Graphite ink (95%)	Al(III)	19-486	DP ASV	[125]
Dimenthylglyoxime (5%) / Graphite	Ni (II)	5	DP CSV	[110]

ink (95%)				
Sumichelate Q10R (8%) / Graphite ink (92%)	Hg(II)	0.0024	DP ASV	[113]

REFERENCES

[1] C. Baird, Environmental Chemistry, W.H. Freeman and company, NewYork (1995).

[2] S. E.Manahan, Environmental chemistry, 6th ëdition, Lewis Publishers, (1994).

[3] L. Barregard, C. Svalander, A. Schutz, G. Westberg, G. Sallsten, Blohm, Molne J, Attman PO, Haglind P, Environ Health Perspect 107(11), Nov 1999, 867.

[4] S. C. Foo, Environmental Research, 60(2) (1993) 267.

[5] O. Lamm, Eur Neurol, 24 (1998) 237.

[6] M. J. Vimy, F. L. Lorscheider, J. Dent Res. 64(8) (1985) 1069.

[7] J. Kawada, J Pharmacobiodyn, 3(3) (1980) 149.

[8] D. L. Smith, South Med J 71 (1978) 904.

[9] T. Tsubaki and K. Irukayama, Minamata Disease: Methylmercury Poisoning, Kodensha, Tokyo, 1977.

[10]P. Kuhnert, Am. J. Obstet and Gynecol,139(1981) 209.

[11]W. D. Kuntz, Am J Obstet and Gynecol. 143(4) (1982) 440.

[12]N. Ghosh, Biomed Environ Sci, 5(3) (1992) 236.

[13]D. Purves, Fundamental Aspects of pollution Control and Environmental Science, 1, ed. Wakeman, Elsevier.

[14]G. A. Drasch, Handbook on Metals in Clinical and Analytical Chemistry (H.G.Seiler, A.Sigel, H.Sigel, eds.), Dekker, New York, 1994, p. 479.

[15]S. Brown, J. Savory, M. R. Wills, Methods in Clinical Chemistry (A.J.Pesce, L.A.Kaplan, eds.), Mosby, St. Louis, 1987, p.405.

[16]S. Chiavarini, C. Cremisini, G. Ingro, R. Morabito, Appl.Organomet.Chem, 8 (1994) 563.

[17]J. C. Gayet, A. Haouz, A. Geloso-Meyer, C. Burstein, Biosens.Bioelectron, 8 (1993) 177.

[18]French standard NF T 90-131 September 1986.

[19]J. Issaq. Hallem and W. L. Zielinski, Jr. Analytical Chemistry.vol.46, N°11 (1974) 1436.

[20]Y. Madrid, C. Cabrera, T. Perez-Corona, and C. Camara, Anal.Chem. 67 (1995) 750.

[21]L. R. Bravo-Sanchez, B. San Vicente de la Riva, J. M. Costa-Fernandez, R. Pereiro, A. Sanz-Medel, Talanta 55(2001) 1071.

[22]C. J. Cappon and J. C. Smith, Ana. Chem, vol. 49, N°3 (1977) 365.

[23]M. J. Vazquez, A. M. Carro, R. A. Lorenzo and R. Cela, Anal. Chem. 69 (1997) 221.

[24]S. Chiavarini, C. Cremisini, G. Ingro and R. Morabito, Applied organometallic chemistry, 8 (1994) 563.

[25]C. F. Harrington, trends in analytical chemistry, 19 (2000) 167.

[26]M. Hempel, H. Hintelman and Rolf-Dieter Wilken, Analyst, 117 (1992) 669.

[27]E. Munaf, H. Haraguchi, D. Ishii, T. Takeuchi, M. Goto, Analytica.Chimica. Acta, 235 (1990) 399.

[28]R. Ritsema and O.F.X.Donard, Applied Organometallic Chemistry, 8 (1994) 571.

[29]P. T. Kissinger and W. R. Heineman, "Laboratory Techniques in Electroanalytical Chemistry" M.Dekker, New York, (1984).

[30]Instruction manual, Dionex Corp, Sunnyvale, USA, (1983).

[31]J. M. Pinilla, L. Hernandes, A. J. Conesa, Anal Chim Acta, 319 (1996) 25.

[32]Yull.D'yachenko, V.V.Kondrat'ev, Zh.Anal.Khim,53 (1998) 401.
[33]Y. Bonfil, M. Brand, E. Kirowa-Eisner, Anal. Chim.Acta, 387 (1999) 85.
[34]M. Kopanica, L. Novotny, Anal. Chim. Acta, 368 (1998) 211.
[35]R. N. Adams, Anal.Chem, 30 (1958)1576.
[36]C. Urbaniczky, K. Lundstrom, J.Electroanal.Chem, 176 (1984) 169.
[37]K. Kalcher, Electroanalysis, 2 (1990) 419.
[38]M. E. Rice, Z. Galus and R. N. Adams, J.Electroanal.Chem, 143 (1983) 89.
[39]N. Motta and A. R. Guadalupe, Anal.Chem,66 (1994) 566.
[40]K. Stulik and V. Pacakova, "Electranalytical measurement in flowing liquids", Ellis Horwood, Chichester, UK, (1987).
[41]J. P. HART, S. A. WRING, Trends Anal Chem; 16 (2) (1997) 89.
[42]M. ALBAREDA-SIRVENT, A. MERKO'I, S. ALEGRET, A review. Sens Actuators B ; 69 (1-2) (2000) 153.
[43]J. Wang, Analyst, 119 (1994) 763.
[44]J-M. Kauffman and J-C. Vire, Anal Chem, 53 (1981) 329.
[45]M. K. Halbert and R. P. Baldwin, Anal. Chem. 57 (1985) 591.
[46]R. P. Baldwin and K. N. Thomsen, Talanta, 38 (1991) 1.
[47]K. Stulik, Pure Appl Chem, 59 (1987) 521.
[48]J. Wang, Anal Chem, 65 (1993) 450R.
[49]E. Wang, H. Ji and W. Hou, Electroanal, 3 (1991) 1.
[50]M. D. Ryan and J. Q. Chambers, Anal. Chem. 64 (1992) 79R.
[51]K. Stulik, Anal.Chim.Acta, 273 (1993) 435.
[52]J-M. Kauffmann, Chim.Nouv, 12, (1994), 1333.
[53]R. W. Murray, "Chemically modified Electrodes" in Electroanalytical Chemistry, A.J.Bard, ed.13, M.Dekker, New York, (1984), 191.
[54]K. Kalcher, J-M. Kauffman, J. Wang, I. Scara, K. Vytras, C. Neuhold and Z. Yang, Electroanalysis, 1(1995) 7.
[55]I. Isildak and A. K. Covington, Electroanal, 5 (1993) 815.
[56]L. D. Whiteley and C. R. Martin, Anal Chem, 59 (1987) 1746.
[57]M. W. Espenscheid, A. R. Ghatak-Roy, R. B. Moore, R. M. Penner, M. N. Szentirmay and C. R. Martin, J.Chem.Soc, Faraday Trans.I., 82 (1986) 1051.
[58]J. Wang and Z. Lu. Anal.Chem. 62 (1990) 826.
[59]E. A. Viltchinskaia, L. L. Zeigman, S. G. Morton, Electroanalysis, 7 (1995) 264.
[60]I. Svancara, M. Matousek, E. Sikora, K. Schachl, K. Kalcher, K. Vytras, Electroanalysis, 9,N°11 (1997) 827.
[61]E. A. Zakharova, V. M. Pichugina, T. P. Tolmacheva, Zh. Anal. Khim, 51 (1996) 1000.
[62]A. M. Vasil'ev, Z. A. Temerdashev, T. G. Tsyupko, Zh Anal.Khim, 54 (1999) 728.
[63]H. Huang, P. K. Dasgupta, Anal.Chim Acta, 380 (1999) 27.
[64]E. A. Zakharova, V. M. Pichugina, N. P. Pikula, Zav Lab, 64 (1998) 9.
[65]S. I. Petrov, L.V. Kukhnikova, Zh. V. Ivanova, Zav Lab, 64 (1998) 13.
[66]I. Svancara, M. Pravda, M. Hvizdalova, K. Vytras, K. Kalcher and Electroanalysis, 6 (1994) 663.
[67]J. P. Hart, S. A. Wring, Electroanalysis, 8 (1996) 617.

[68]G. Sittampalam, G. S. Wilson, Anal. Chem. 55 (1983) 1608.
[69]J. Wang, L. D. Hutchins. Anal. Chem. 57 (1985) 1536.
[70]L. S. Kuhn, S. G. Weber, K. Z. Ismail, Anal. Chem. 61 (1989) 303.
[71]J. Wang, L. D. Hutchins-Kumar, Anal. Chem. 58 (1986) 402.
[72]B. Hoyer, N. Jensen, Talanta 42 (1995) 767.
[73]C. M. A. Brett, D. A. Fungaro, J. M. Morgado, M. H. Gil, J. Electroanal. Chem. 468 (1999) 26.
[74]C. M. A. Brett, D. A. Fungaro, J. Braz. Chem. Soc, Vol. 11 (2000) 3.
[75]C. M. A. Brett, V. A. Alves, D. A. Fungaro, Electroanalysis 13 (2001)212.
[76]Z. Chen, Z. Pourabedi, D. B. Hibbert, Electroanalysis, 11 (1999) 964.
[77]Zh. Hu, C. J. Seliskar, W. R. Heineman, Anal.Chim.Acta, 369 (1998) 93.
[78]F-M. Matysik, P. Glaser, G. Werner, Fresenius J.Anal.Chem. 349 (1994) 646.
[79]P. Ugo, L. M. moretto, P. Bertoncello, J. Wang, Electroanalysis, 10 (1998) 1017.
[80]J-M. Zen, J-W. Wu, Anal. Chem. 68 (1996) 3966.
[81]J-M. Zen, M-J. Chung, Anal Chem, 67 (1995) 3571.
[82]J-M. Zen, J-W. Wu, Electroanalysis, 9 (1997) 302.
[83]A. Witkowski and A. Brajter-Toth, Anal.Chem, 64 (1992)635.
[84]F-Y. Song, K-K.Shiu, Journal of electroanalytical Chemistry 498 (2001) 161.
[85]J-M. Zen, F. S. Hsu, N. Y. Chi, S. Y. Huang, M. J. Chung, Anal. Chim.Acta, 310 (1995) 407.
[86]K.Wagner, J.W.Strojek, K.Koziel, Analytica Chimica Acta 447 (2001) 11.
[87]P. Ugo, L.M. Moretto, G.A. Mozzocchin, Anal.Chim.Acta, 273 (1993) 229.
[88]P. Ugo, L. Sperni, L. M. Moretto, Electroanalysis, 9 (1997) 1153.
[89]I. Turyan, D. Mandler, Fresenius J.Anal.Chem.,349 (1994) 491.
[90]B.J. Palys, M. Skompska, K. Jackowska, J. Electroanal. Chem. 433 (1997) 41.
[91]Q. Cai, S. B. Khoo, Anal.Chem, 66 (1994) 4543.
[92]Kh. Z. Brainina, N. A. Malakhova, N. Yu. Stojko, Fresenius J. Anal. Chem. 368 (2000) 307.
[93]S. I. Petrov, L.V.Kukhnikova, Zh.V. Ivanova, ZavLab, 64 (1998) 6.
[94]L. D. Svintsova, N. N. Chernyshova, ZH. Anal.Khim. 52 (1997) 917.
[95]E. E. Fomintseva, E. A. Zakharova, N. P. Pikula, Zh. Anal. Khim, 52 (1997) 657.
[96]K. Kalcher, Electroanalysis, 2 (1990) 419.
[97]I. Svancara, K.Vytras, J. Barek, J. Zima, Critical reviews in Analytical Chemistry, 31(4) (2001) 311.
[98]Z-Q. Zhang, H. Liu, H. Zhang , Y-F. Li, Anal.Chim.Acta. 333 (1996) 119.
[99]S-S. Huang, Y-D. Cheng, B-F. Li, G-D. Liu, Mikrochi. Acta. 130 (1998) 97.
[100] K. Kalcher, J-M. Kauffman, J. Wang, I. Svancara, K. Vytras, C. Neuhold, Z. Yang, Electroanalysis, 7 (1995)5.
[101] R. Metelka, K. Vytras, A. Bobrowski, J. Solid State Electrochem, 4 (2000) 348.
[102] R. Martinez, M. Teresa Ramirez, I. Gonzalez, Electroanalysis, 10 (1998) 336.
[103] K. Ravinchandran, R.P. Baldwin, J. Electranal.Chem. 126 (1981) 293.
[104] Q.J. Chi, W. Goepel, T. Ruzgas, L. Gorton, P. Heiduschka, Electroanalysis, 9 (1997) 357.
[105] J. Wang, B. Greene, and C. Morgan, Anal.Chim.Acta, 158(1984)15.

[106] H. Kahlert, F. Scholz, Electroanalysis, 9 (1997) 922.
[107] S. B. Khoo, J. Zhu, Analyst, 121 (1996) 1983.
[108] Q. Cai, S.B. Khoo, Electroanalysis, 7 (1995) 379.
[109] M. Guttman, K-HL. Beyer, Fresenius J.Anal.Chem, 356 (1996) 263.
[110] J. Wang, V.B. Nascimento, J. Lu, D.S. Park, L. Angnes, Electroanalysis, 8 (1996) 635.
[111] I. Helms, F. Scholz, Fresenius J.Anal.Chem, 356 (1996) 237.
[112] J. Labuda, H. Kogova, M.Vanickova, Anal.Chim.Acta, 305 (1995) 42.
[113] P. Ugo, L.M. Moretto, P. Bertoncello, J. Wang, Electroanalysis, 10 (1998) 1017.
[114] P. Kula, Z. Navratilova, P. Kulova, M. Kotoucek, Anal.Chim.Acta, 385 (1999) 91.
[115] Ch. Wang, H. Zhang, Y. Sun, H. Li, Anal.Chim.Acta, 361 (1998) 133.
[116] Ch.Wang, H. Li, Electroanalysis, 10 (1998) 44.
[117] E-D. Jeong, M-S. Won, Y-B. Sbim, Electroanalysis, 6 (1994) 887.
[118] A. Walcarius, C. Despas, P. Trens, M.J. Hudson, J. Bessiere, J.Electroanal.Chem.453 (1998) 249.
[119] H. Yao, G.Ramelow, J.Talanta. 45 (1998)1139.
[120] Z-U. Bae, Y-L. Kim, H-Y. Chang, F. Song, Anal.Sci.Technol.8 (1995) 611.
[121] S.B. Khoo, Q. Cai, Electroanalysis, 8 (1996) 549.
[122] M-S. Woon, D-W. Moon, Y-B. Shim, Electroanalysis, 7 (1995) 1171.
[123] T. Molina-Holgado, J.M. Pinilla-Macias, L. Hernandez- Hernandez, Anal.Chim.Acta, 309 (1995) 117.
[124] R. Agraz, J. de.Miguel, M-T. Sevilla, L. Hernandez, Electroanalysis, 8 (1996) 565.
[125] P. Akhtar, H.A. Devereaux, A.J. Downard, B. O'Sullivan, K.J. Powell, Anal. Chim. Acta, 381 (1999) 49.

Chapter II: Electrochemical methods used

In this chapter, we will review the theoretical and practical foundations of the electrochemical techniques used in this work.

I- AMPEROMETRY

Ampëromëetry is a technique which consists of determining the intensity of the current flowing through a ëlectrochemical cell when the working electrode is subjected to a potential allowing ëlectrochemical transfer. A conventional three-ëlectrode system is generally used: indicator electrode, auxiliary electrode and reference electrode.

Quantitative ampërometric analyses consist of rëfërering to a calibration line or using the dosës additions technique. These methods verërimentally express the proportionality coefficient between the diffusion limit current and the concentration of the species ëко^^^E The measurements are carried out under well-defined and reproducible conditions (nature and concentration of the electrolyte, pH, potential, agitation).

In amperometry, the residual faradic current due to the presence of traces of ëlectroactive impurities limits the sensitivity of the analysis. The capacitive current resulting from switching on the electrode is very quickly cancelled out. The sensitivity limit is approximately 10^{-8} M. The residual current also depends on the nature of the electrode and the chemical or electrochemical treatments that have been applied to it [1-2].

II- CYCLIC VOLTAMETRY

Cyclic voltamëtrie has ël ë ëtudiëe tlK'oriquement antërieurement by several authors [3-7]. This technique is based, like any ëlectrochemical ampërometric technique, on the exchange of electrons between a ëк^re^ and a chemical spëce in solution (solution called ëlectrolyte) or on the surface of the electrode. A chemical molecule

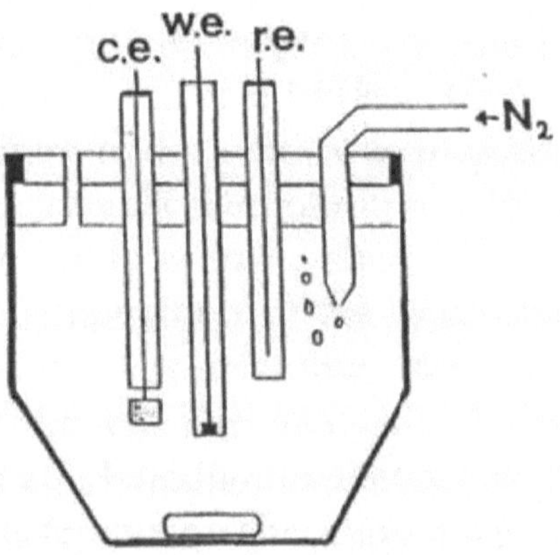

Figure 1: Schematic diagram of the ëlectrochemical cell. **w.e.**: working electrode, **c.e.:** auxiliary electrode and **r.e.**: reference electrode.

Having the capacity to give up or to capture one or more electrons is qualified as an electroactive species and can be present either in an oxidised form (Ox) or in a reduced form (Red) according to the following fundamental equilibrium:

$$\text{Red} \underset{(2)}{\overset{(1)}{\rightleftharpoons}} \text{Ox} + n\,e^- \qquad (a)$$

For a reversible redox couple, characterised by very rapid charge transfer, the concentrations of Ox and Red species in the vicinity of the electrode, Cox and Cred respectively, are related to the electrode potential, E, by Nernst's law:

$$E = E^\circ + \frac{RT}{nF} \ln \frac{Cox}{Cred}$$

WhereE°: the standard potential of the Ox/Red pair,
R: the perfect gas constant,
T: the tempërature in Kelvin, n: the number of Electrons involved, F: Faraday's constant.
The existence of a ëlectrochemical reaction implies the transport of the substance ёк^та^^ consumëe a lklectrode, from the solution to the surface where it undergoes the transfer! ëlectronique. This transport of material automatically results in the appearance of a concentration gradient, which causes a diffusion flow of the substance. The layer of solution near the electrode in which concentration gradients exist is called the diffusion layer.
Voltamëtrie li^aire is a particularly inkressante ëlectrochemical technique for obtaining qualitative and quantitative information about an ëlectrochemical reaction [8,9]. It is a simple, easy-to-use method that provides a wealth of information about the processes taking place at the electrode surface (adsorption, passivation, etc.) and the behaviour of the electroactive species detected (fast, reversible system, number of electrons exchanged, etc.). The method consists of varying the potential of a stationary working electrode liiK'airement in time. During this sweep, the corresponding current is recorded. An alkr-return potential sweep between two potential values dëfinishes the cyclic li^aire voltamëtrie, commuiK'inent called cyclic voltamëtrie.
A voltamogram I = f(E) obtained on a stationary ëlectrode for a reversible Ox/Red redox couple is shown in figure 2, considering that only the Red form is initially present in solution and does not adsorb to the electrode surface as a function of time. The initial potential (E_i) is chosen so that the current Hë at the oxidation of Red is iK'gligeable ($E_i <$ E°: standard potential of the redox couple). The current measured at this potential corresponds to the capacitive current. By sweeping towards positive potentials, the first пюШё of the cycle i = f(E) can be recorded and presents a peak of anodic current asso^ to direction 1 of reaction (a).
The cara^ristic appearance of the peaks of a voltamogram recorded on a stationary ëlectrode can be explained by the variation in the concentrations of Ox and Red within the diffusion layer in the vicinity of the electrode surface. Indeed, when the potential becomes more positive, and in particular, when it approaches E°, the a^teration of the oxidation process from Red to Ox results in an increase in the anodic current. The higher the potential, the greater the rate of transformation of Red into Ox and the greater the depletion of Red in the vicinity of Electrode. From a potential of

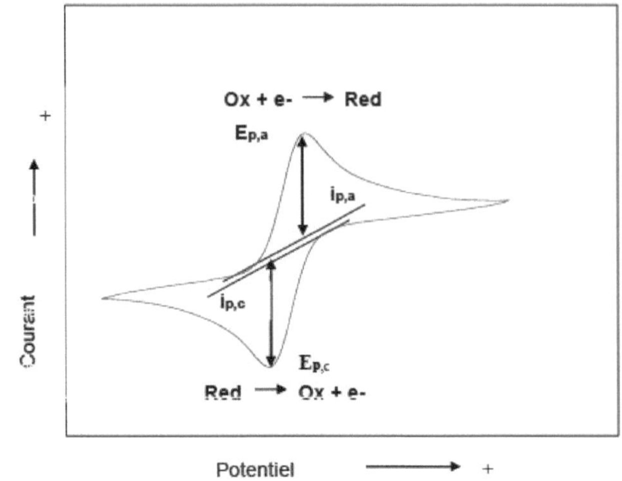

Figure 2. Cyclic voltamogram recorded for a single Red species in solution

Appelë anodic peak potential ($E_{p,a}$), 1 depletion is such that it results in a dëgrowth of the anodic current. During the return sweep, the Ox molecules generated during the first part of the cycle and present in the diffusion layer in large quantities, are then reduced and a cathodic peak characterised by $I_{p,c}$ and $E_{p,c}$ is observed.

The voltamogram obtained is characterised by several parameters. In the case of a reversible system. The peak currents $i_{p,a}$ and $i_{p,c}$ are identical, proportional to the concentration of ëelectroactive species. They increase with the square root of the scanning speed according to the following equation:

$$I_p = 2.69 \times 10^5 \times n^{3/2} A D^{1/2} v^{1/2} C \qquad \text{(à 25°C)}$$

With

D: diffusion coefficient of the species ëtudiëe (cm^2 .s),$^{-1}$

v: sweep speed (V.s),$^{-1}$

A: surface area of the electrode (cm),^{2}

C: concentration of the ëtudiëe species (mol.cm).$^{-3}$

The position of the peaks on the axis is a function of the value of the standard redox potential. For a reversible system, the standard potential is given by :

$$E° = (E_{p,a} + E_{p,c}) / 2$$

The separation between the two peaks is given by :

$$\Delta E_p = E_{p,a} - E_{p,c} = 0.058 / n \qquad \text{(en Volts)}$$

(in Volts)

n: the number of electrons involved.

As soon as the rate of electron transfer is lower than the rate of matter transfer, the system is said to be 'electrochemically irreversible'. For such a system, Nernst's law is no longer applicable at the surface of the electrode, and the expressions for the peak currents and potentials, which are not presented here, include additional parameters (transfer coefficient, heterogeneous rate constant).

III- PULSE TECHNIQUES

In general, voltamperometry is an electroanalytical method in which information about the analyte is obtained from the measurement of current as a function of voltage at the working electrode. This method is used at a fundamental level to understand the mechanisms of oxidation and reduction reactions in different media and to study adsorption and electron transfer processes on chemically modified surfaces. In the early 60s, a number of developments were made to improve the sensitivity and selectivity of these voltamperometric methods. In fact, the presence of a non-negligible charge current, linked to the double layer represented by the electrode-solution interface, considerably limits the application of this technique for analysis. The capacitive current, resulting from the behaviour of the double layer as a capacitor, is very difficult to eliminate.

A number of solutions were put forward thanks to advances in electronics and the development of suitable detection circuits. To differentiate between faradic current (directly linked to electron transfer) and capacitive current (linked to the double layer represented by the electrode-solution interface), various signals were superimposed on the DC voltage normally applied. These techniques, known as alternating, pulsed and differential pulsed, were first used to develop the polarographic technique and then applied to voltamperometric methods. The gain in detection limit is very appreciable (by a factor of 10^2 and 10).[3]

The first impulse polarographic method was developed by Barker [10,11] in 1957. But it is to R. and J.Osteryoung [12,21] and their collaborators that we owe the major part of the dëveloppement of the whole range of impulse techniques. There are two groups of current pulse voltampërometric techniques:

> voltampëromëtrie impulsionnelle differentielle (or constant pulse): on superimposes a squared periodic pulse on a linear "sweep ramp".

> Square voltage voltammetry

III.1 Differential drive voltametry

The principle of this method [22-25] is to superimpose pulses of constant amplitude on a DC voltage which varies linearly with time. In addition to the DC faradic component, the resulting current has a rapidly decreasing capacitive component and a faradic component due to the pulse. The current is measured twice: the first time before the pulse is given, i.e. any current not due to the pulse is measured and stored in memory; the second time at the end of the pulse, so that the capacitive current due to the pulse has had time to cancel out. The current transmitted is the difference between the two measurements ***ЛI = I(2) - I(1)***. This principle eliminates the capacitive component and the direct current. The detection limit reached is of the order of ppb.

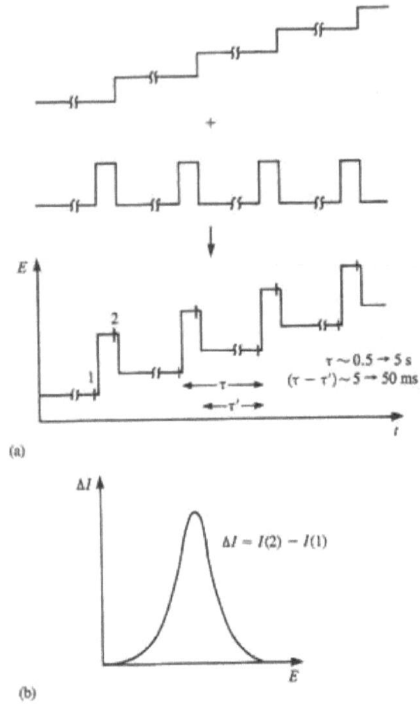

Figure 3 : Differential pulse voltammetry (a) superposition of pulses (b) the profit of the current transmitted as a function of the potential

III.2 Square wave voltametry

The excitation signal is obtained by superimposing on a potential "staircase", of height AEp, a symmëtrical square wave of amplitude AEs in phase with the frequency of the "staircase steps" (Figure 4a). The current is ëchantIIIonnë for very short time intervals at the end of each pulse where the load current is virtually constant. The resulting voltamogram (Figure 4b) actually corresponds to the diffërence of the anode and cathode currents ëchantillonnës (Ai = ii - ii), ë^тят: thus the capacitive current. The peak height is proportional to the concentration of spëce ëlectroactive, and the peak potential corresponds to the half-wave potential observed in normal polarography.

The potential step AEp, the përiode of ***m*** and the pulse amplitude AEs are three parameters that can influence the peak width and height, i.e. the resolving power and sensitivity^ of the mëthod.

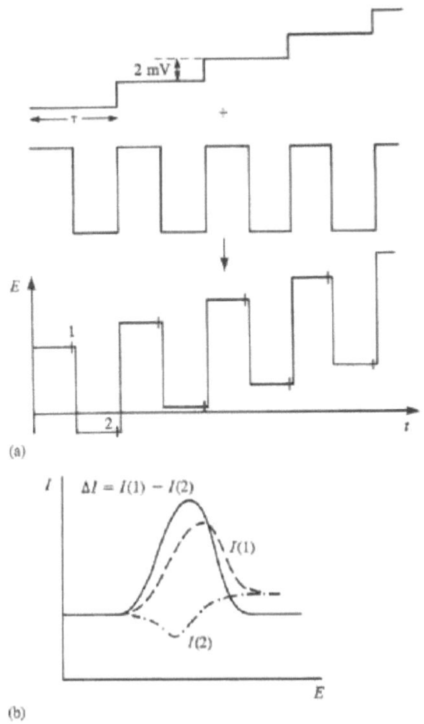

Figure 4: (a) Imposed voltage versus time. Assembly of a stepped voltage and a symmetrical square-wave voltage. **(b)** Voltampërogram resulting from the of a "square wave" pulse

IV- VOLTAMETRY BY ANODIC REDISSOLUTION

IV-1 Introduction

Although the procedure itself has been known for a long time [22-29], voltamëtric analysis by redissolution has seen considerable growth in recent years due to the very high sensitivities required by certain analyses in the biological, chemical, environmental and electronics fields [30]. This method allows several elements to be determined simultaneously at concentrations in the ppb range. The technique can also be used in complex environments and offers excellent precision. Finally, the instrumentation used is inexpensive.

Voltamëtrie by anodic redissolution is particularly applicable to the determination of metals, although several organic compounds have been analysed. The media studied are highly variable. Drinking water, natural water, rainwater and marine water have frequently been analysed.

The general principles governing this technique are outlined below.

IV-2 General principles

This method (Anodic Stripping Voltammetry ASV) consists of reducing the cation in question to a cathodic potential determined using an electrode chosen according to the analysis to be carried out. After a rest period, the potential is swept towards less negative values and the deposited metal or amalgam redissolves when the potential corresponds to its reoxidation. The resulting current is measured.

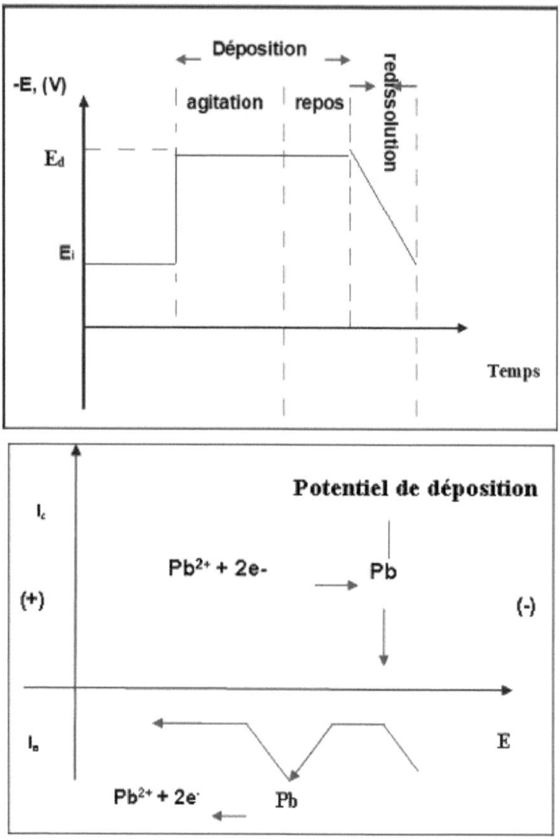

Figure 5: The ëtages of voltammetric measurement by anodic redissolution.

a- Concentration stage

Three main factors need to be taken into account in this process

- Choice of ëlectrodëposition potential

The ëlectrodëposition potential must be chosen so as to ensure the reduction of the cation while ëavoiding dëposing other substances which may interfere.

- Electroposition time

In general, the metal concentration in the amalgam is proportional to the electrolysis time. This time will therefore be adapted to the quantity of ions to be determined. It is generally accepted that a time of two minutes is appropriate for a 10^{-6} M solution; this time can be extended to sixty minutes if the concentration is less than 10^{-9} M.

- Mass transport

An increase in the electrolysis current, and hence an increase in the amount of metal deposited or amalgamated, can be achieved by stirring the solution. Convection-phënomënes will thus be added to the diffusion process and increase the supply of matiëre. An increase in the surface area of the electron also makes it possible to sensitise the technique without altering the operation time.

b-Rest time

The rest period observed between the concentration phase and redissolution favours homogeneous distribution of the amalgam. This period also ensures stabilisation of the solution in the cell. This resting time must be kept constant over a series of measurements.

c-Redissolution stage

The redissolution phase cannot always be carried out in the medium in which the metal concentration was achieved, and it is then necessary to change the supporting electrolyte after the rest phase or to add an additional compound to the supporting electrolyte after the rest phase or to add an additional compound to the initial electrolyte. This procëdë is to be considered when the peak potentials of two mëtals are too close to each other.

The dë sensitivity depends mainly on the potential sweep rate, which is generally set at between 5 and 100mV/s. However, an increase in this parameter also generates a non-negligible capacitive current which adversely affects the definition of the baseline by giving a significant residual current.

V- REDISSOLUTION POTENTIOMETRY

This technique, which was recommended towards the end of the 1970s, has benefited greatly from advances in computers and microprocessors, allowing rapid control and recording of data [32,33]. PSA is similar to ASV in that the accumulation procedure is the same, the only difference being that the redissolution step is carried out in the presence of an oxidant in the solution or alternatively by an applied anodic current which ensures the oxidation of the reduced analytes.

It is based on an electrolytic step, designed to preconcentrate the substance to be analysed on the surface of the electrode where a reduction potential is applied, followed by measurement of the change in potential of the working electrode as a function of time during the redissolution step.

Preconcentration stage (electrolytic): $Mn+ + ne- \rightarrow M(Hg)$.

Redissolution stage (chemical) : $M(Hg) + oxidant \rightarrow Mn^{+} + ne-$

(oxidant = dissolved oxygen or mercuric ions).

This technique has the same advantages as SQASV. The dëaëration of ëcЬaп1Шoп8 is, gënërally, unnecessary since dissolved Гoxyдёпс can be ийHзё as an oxidant. Measurement of the Mectrode potential as a function of time gives a voltamogram dt/dE as a function of time which provides both quantitative and qualitative information on the mëtals present in the solution.

REFERENCES

[1] B. D. Epstein, E. Dalle-Molle and J. S. Mattson, Carbon, 9 (1971) 609.
[2] V. L. Snoeyink and W. J. Weber, Progr. Surface Membrane Sci. 5 (1972) 63.
[3] W. Kemula and Z. Kublik, Nature, 182 (1958) 793.
[4] R. S. Nicholson, Anal.Chem.37 (1965) 667.
[5] A. J. Bard and L. R. Faulkner, "Electrochimie, principes, mëthodes et applications" Masson, Paris, (1983).
[6] R. N. Adams, "Electrochemistry at solid electrodes" M.Dekker, New York, (1969).
[7] R. S. Nicholson and I. Shain, Anal. Chem. 36 (1964), 706.
[8] P.T. Kissinger and W.R. Heineman, "Laboratory Techniques in Electroanalytical Chemistry" M.Dekker, New York, (1984).
[9] D. Bauer, M. Lamache, C. Colin and G. Cote, "Voltampëromëtrie sur ëlectrode solide", Technique de l'ingënieur, p.10, (1984),2125.
[10]G. C. Barker, A. W. Gardner, Proc.Congr.Mod.Anal.Chem.Ind, St Andrews (1991) 957.
[11]G. C. Gardner, A. W.Gardner, Z. Anal. Chem. 173 (1960) 79.
[12]E. P. Parry, R. A. Osteryoung, Anal. Chem. 36(1964) 1366.
[13]E. P. Parry, R. A. Osteryoung, Anal. Chem. 37(1965) 1634.
[14]H. E. Keller, R. A. Osteryoung, Anal. Chem. 43 (1971) 342.
[15]J. Osteryoung, R. A. Osteryoung, Amer.Lab., 4 (1972) 8.
[16]J. H. Christie, R. A. Osteryoung, J. Electroanal. Chem. 49 (1974) 301.
[17]D. J. Myers, R. A. Osteryoung, J. Osteryoung, Anal. Chem. 46 (1974) 2089.
[18]J. Osteryoung, R. A. Osteryoung, Anal. Chem, 57 (1985) 101A.
[19]J.J. O'Dea, J. Osteryoung, "Electroanalysis Chemistry", vol 14, Ed. A.J.Bard, Dekker, N.Y.,1986.
[20]Z. Stojek, J. Osteryoung, Anal. Chem. 63 (1991) 839.
[21]J. Osteryoung, Acc. Chem. Res. 26 (1993) 77.
[22]AMEL, Modern Polarographic and voltametric Analysis A Handbook, version 3. July 1994.
[23]J. Wang, Analytical Electrochemistry, second edition, Wiley-VCH (2000).
[24]C. M. A. Brett, A. M. O. Brett, Electranalysis, Oxford University Press Inc. New York 1998.
[25]J. Wang, "Stripping Analysis, Principles, Instrumentation, and Applications", V.C.H publishers Inc, New York, (1985).
[26]E. Barendrecht, "Stripping Voltammetry", in A.J.Bard, "Electroanalytical Chemistry", M. Dekker, New York, V.11, (1967).
[27]T. R. Copeland, R. K. Skogerboe, Anal.Chem, 46 (1974) 1257A.
[28]I. Shain, "Stripping Analysis", in I.M.Kolhoff and P.Elving, "Treatise on AnalyticalChemistry", part I, Interscience, New York, (1963), Vol IV.
[29]F. Vydra, K. Stulik, E. Julakova, "Electrochemical Stripping Analysis", Ellis Horwood Ltd, Chichester, (1976).
[30]A. M.Bond, "Modern Polarographic method in Analytical Chemistry", M.Dekker, New York, (1980), Chap9.
[31]D. Jagner, Analyst, 107 (1982) 593.
[32]A. M. Graabaek, B. Jeberg, intern.Lab., 9 (1982) 33.

EXPERIMENTAL PART

Chapter III

AMPEROMETRIC METHOD FOR DETERMINING TRACES OF MERCURY (II) BY THE FORMATION OF A COMPLEX WITH L-TYROSINE.

I- INTRODUCTION

Electroanalytical methods using chemically modified electrodes for the determination of mercury have undergone considerable development in recent years. The first modifications involved gold electrodes [1-4], these electrodes have traditionally been used for mercury determination because of their high sensitivity. Platinum, graphite and carbon paste electrodes have been used as substrates for the gold film for mercury detection [5-11]. Mercury forms an amalgam with the gold on the surface of these electrodes, enabling it to be detected.

However, the main disadvantage of using these gold electrodes in the analytical field is the regeneration of the electrode surface and the slow and complicated electrochemical pre-treatment required each time to achieve good reproducibility.

Other work has focused on the use of carbon paste electrodes (CPE). Since 1964 [12], several studies aimed at the determination of mercury have been published on modified carbon paste electrodes. The carbon paste has been modified with various organic compounds selective for mercury [13-17]. Ion exchangers such as Amberlite or HYPHAN have also been used [18,19]. Ugo and colleagues modified screen-printed carbon electrodes with Sumichelate Q10R (containing dithiocarbamate groups) for the analysis of traces of mercury; this modification enabled a very low detection limit of the order of 12pM to be achieved [20].

The use of other mercury-selective compounds as modifiers of solid electrodes has been extensively studied in the literature. The application of certain ion exchangers to the development of new voltammetric ion exchange methods for the selective determination of mercury has recently been published. Moretto et al have used Tosflex, a perfluorinated anion exchanger, to modify the glassy carbon electrode [21].

The sëlectivitë has ële markedly improved by introducing chelating agents within a polyтёre. Thus, Chen et al mixed with Nafion, 1 diethyl- dithiocarbamic acid (compound I) and 18-courrone ether-6 (compound II). They were able to analyse Pb^{2+} , Cu^{2+} and Hg^{2+} using DPASV. The concentration determined ranged from 5.10^{-7} to 5.10^{-5} M using compound I and from 2.10^{-6} to 4.IO-5M for compound II [22]. Other anion-exchange polymëers such as poly(4-vinylpyridine) [23] and a cationic polypyrrole derivative [24] have also been used for mercury analysis. A study by turyan et al concerning the modification of the glassy carbon electrode by "Kryptofix-222" achieved a detection limit of less than 0.2gg/l using SQWASV[25]. The same author recently used a permselective Kryptofix-222 membrane in which poly(4-vinylpyridine) was incorporated for mercury analysis [26].

All the determination methods described above are based either on the complexation of heavy metals by selective compounds or on ion exchange.

In this chapter, we have used the amino acid L-tyrosine for the determination of mercury.

This determination is based on the inhibition of the tyrosine response in the presence of mercury.
Several metal ions form complexes with amino acids and small peptides. This complexation has been extensively studied to better understand the phenomëne of transition metal complexation at the protein level. Recently, the affinities of lithium, sodium, copper [27] and zinc [28] ions with the most common amino acids have been measured. The complexes formed between certain cations and tyrosine have also been examined. Ryzhov et al studied cation-л interactions in the complexation of Na^+ and K^+ cations with the aromatic amino acids Phe, Tyr and Trp [29]. This type of non-covalent bond is very common between proteins and their cationic substrates. This is also the case for bis(tyrosinato) copper(II) [30], bis(tyrosinato) palladium(II) [31] and (glycine-leucine-tyrosine) Cu(II)[32] crystals.
Remember that L-tyrosine is a aттë acid that occurs naturally in the body.
It has a pKa of 2.2 (-COOH group) and 9.1 ($-NH^{3+}$ group) and its formula is shown below.

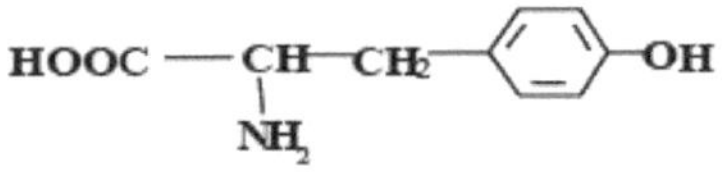

L-Tyrosine

Tyrosine is important for the structure of almost all protëines in the body [33]. It ës also a direct precursor of several neurotransmitters [34-35], including L- dopa, dopamine, nopëpinëphrine and l^pinephrine.

PART II

II- EXPERIMENTAL PART

II-1 Equipment

Ampometric measurements were ëlë rëalisëtric using a BAS CV-27 instrument coupled to a BAS LC-3D electrochemical detector and an X-Y recorder. Cyclic voltammetry was ëlë rëalisëtric using an Autolab PGSTAT 10 potentiostat-galvanostat.

Ampërométric measurements were ëlë made under constant stirring at 300 rpm.

The ëlectrochemical cell used is a conventional three ëlectrode cell. The rëfëference electrode is Ag/AgCl ([KCl]=3M), the auxiliary electrode is a platinum wire and the working electrode is a platinum ëlectrode with a gëomëtrical surface area of 3.14 mm .[2]

II-2 Reagents

The chemicals used are all of "for analysis" quality. The water used to prepare all the solutions is bi-distilled water. The mercury stock solution is a 10^{-3} mol l solution^{-1} Hg^{2+} prepared from HgCh.

Sodium phosphate solutions of pH 2 to 11 were used in this work. Solutions of pH 2 to 5 are prepared from monobasic phosphate and adjusted with phosphoric acid. Solutions at pH 5 to 9 were prepared from 0.1 mol l^{-1} monobasic and dibasic sodium phosphate solutions. Solutions of pH 10 and 11 are prepared from dibasic phosphate adjusted with sodium hydroxide.

The amino acids L-tyrosine, L-cysteine hydrochloric acid, L-tryptophan, and the molecules L-dopa (3,4-dihydroxyphenylalanine), dopac (3,4-dihydroxyphenylacetic acid) and dopamine (3-hydroxytyramine) are products marketed by Aldrich.

Methylmercury chloride and phenylmercury acetate come respectively from "JM Alfa Products, Karlsruhe, F.R.Germany" and "Sigma".

All measurements are taken at room temperature, 25 ± 1°C.

III- RESULTS AND DISCUSSION

III-1 Electrochemical oxidation of L-Tyrosine

The amino acid L-tyrosine oxidises on the platinum electrode at positive and negative potentials.

gives a weakly defined electrochemical response [36-39], The mechanism of its electrochemical oxidation given in the literature is as follows:

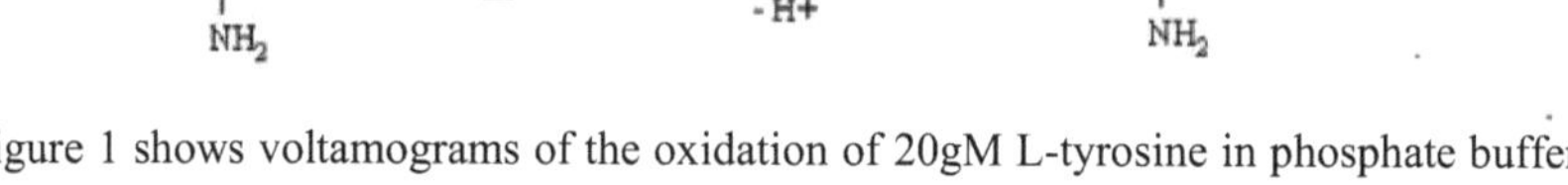

Figure 1 shows voltamograms of the oxidation of 20gM L-tyrosine in phosphate buffer at pH=7 in the absence and presence of mercury.

The cyclic voltametry of the tyrosine solution alone shows an irreversible oxidation wave at E=+0.8V. This result is in agreement with the work of kenneth et al [40] who ëtudië the oxidation of tyrosine on a Pt ëlectrode, the irreversible oxidation peak at ële obtained at +0.8V using a 3mM solution of tyrosine at pH=6.5 . The addition of an aliquot of a mercury solution leading to a concentration of 50gmoll'ı in the ëlectrochemical cell produces a 20% decrease in peak current and a dëcalage of the peak towards the more ^gative potentials. The decrease in current is probably explained by the formation of a complex between tyrosine and mercury ions.

In order to better understand this phënomëne between tyrosine and mercury, we have used ampërométrie as a technique. This ëtude consists in ëstudying the effect of the presence of mercury on the tyrosine signal.

III-2 Optimising the electrochemical response of tyrosine

III-2-1 Influence of working potential :

The choice of working potential is necessary for the ampërométric iritude we want to achieve. Preliminary studies show that L-tyrosine oxidises at positive potentials in a neutral medium. We therefore ëtudië the response of tyrosine at pH7 for each value of potential applied. The ëtudië range is between 0 and IV/Ag/AgCl. The tyrosine oxidation currents obtained for each applied potential are shown in Figure 2.

We observe that tyrosine oxidation occurs from a potential greater than 0.6V.

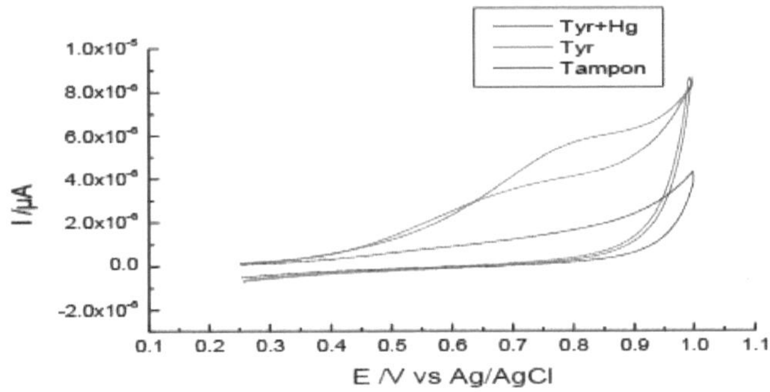

Figure 1: Voltamograms of the electrooxidation of 20ııM L-tyrosine (red) and 20ııM L-tyrosine + 50 LIVI Hg^{2+} (blue) in 0.1 M phosphate buffer at pH 7 (black). Scanning speed 50 mV s .$^{-1}$

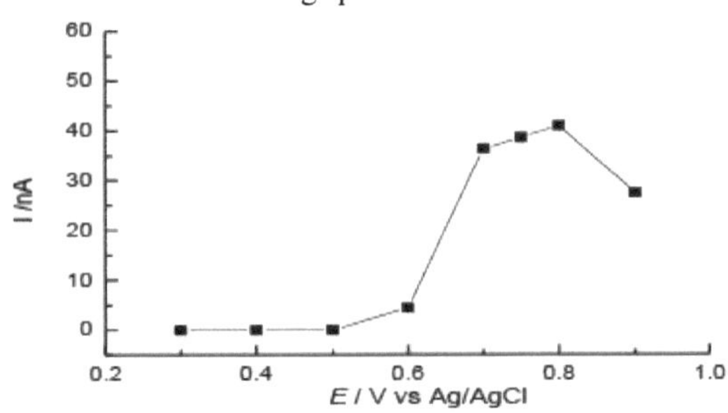

Figure 2: The effect of applied potential on the oxidation current of 20ııM L-tyrosine in 0.1M phosphate buffer at pH=7.

Earlier studies by the Ogura team showed that this potential value corresponds to the complementarily flat orientation of the tyrosine benzene ring on the electrode surface [41]. Adsorption of amino acids onto a platinum ëlectrode, in basic solutions, is accomplished through the terminal carboxylic group which is in the anionic form and begins to adsorb from +0.4V/ Ag/AgCl. The benzene ring begins to be oriented parallel to the electrode surface from +0.3V and the flat orientation is completed at approximately +0.6V as shown in the diagram below.

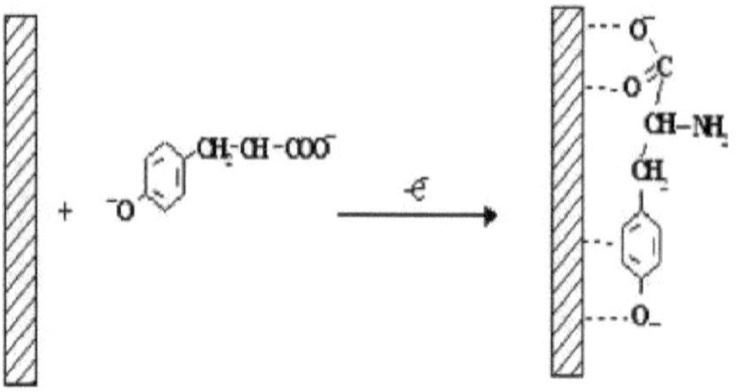

Similar studies on silver [42] and gold [43] electrodes gave the same results. Adsorption begins via the ionised carboxylic groups and then the aromatic ring is oriented parallel to the surface.

III-2-2 Influence of pH

A pH-dependent study was carried out to determine the influence of pH variation on the tyrosine oxidation signal. Amperometric measurements of tyrosine oxidation were performed at pH values ranging from 2 to 11. Figure 3 shows that the tyrosine oxidation signal is pH dependent. The best response is obtained in the pH range between 7 and 9.

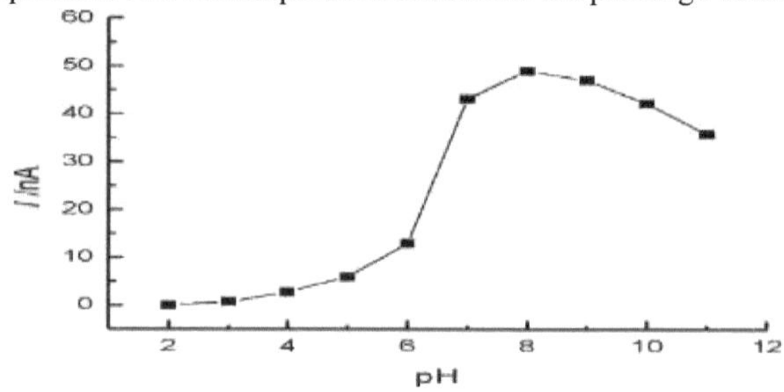

Figure. 3: The effect of pH on the oxidation current of 20μM L-tyrosine at +0.75 V vs. Ag/AgCl.

Given that the best results, in terms of current, are obtained from potentials greater than 0.6V and in a pH range between 7 and 9, we will continue our work by studying the effect of mercury addition on the tyrosine oxidation signal.

III-3 Procedure for determining mercury (II)

The platinum electrode is immersed in a cell containing 10 ml of phosphate buffer, and once the residual current has stabilised, tyrosine is injected into the cell. Figure 4 shows typical current-time curves. The oxidation current of the added tyrosine is measured as ***I1***. When mercury ions are added to the solution, they form a non-electroactive complex with tyrosine, a decrease in the tyrosine oxidation current is observed, the ***I2*** current is measured after 11min. The percentage decrease in the tyrosine signal due to the addition of Hg ions^{2+} can be expressed as follows:

$$I\% = \frac{I_1 - I_2}{I_1} * 100$$

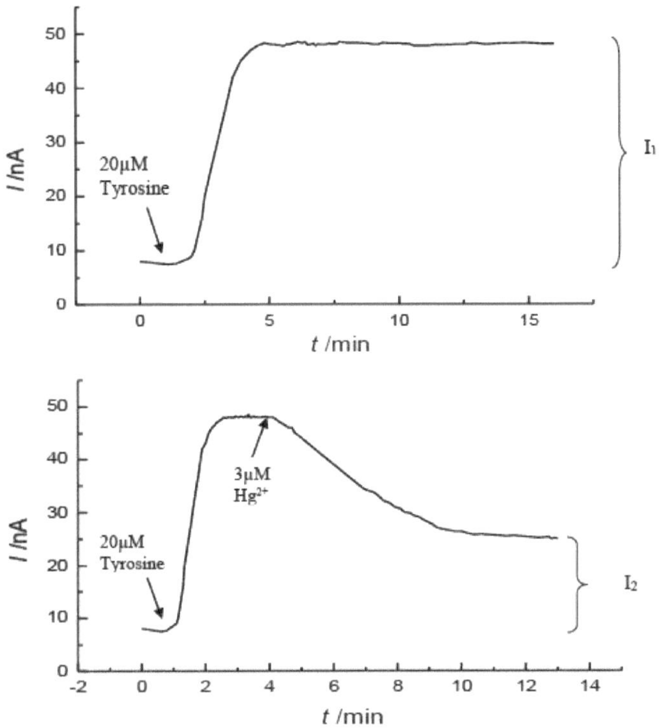

Figure 4: Typical current-time curves obtained before and after the addition of mercury ions at E= +0.75V and pH= 7.

Typical current-time curves obtained before and after the addition of mercury ions has E= +0.75V and pH= 7.

III-4 Effect of tyrosine concentration

The influence of tyrosine concentration has ële ëtudiëe. To this end, different solutions with concentrations of 4, 20, 50 and 100 Limol l^{-1} were prepared. Figure 5a shows that by increasing the concentration of tyrosine, the oxidation current increases.

We note that a high tyrosine concentration of 100 LIVI gives an unstable signal. This is probably due to the orientation of the tyrosine molecules on the electrode surface. It has been shown that the orientation of adsorbed molecules depends on the degree of surface coverage [44].

Addition of the same concentration of mercury ILIVI to solutions of different tyrosine

concentrations shows that the rate of inhibition does not vary. Figure 5b shows the rate of inhibition of the tyrosine signal for the different concentrations. In fact, whatever the tyrosine concentration used, the inhibition rate recorded for a concentration of 3gM of Hg^{2+} is equal to 30%.
In view of the results obtained, a concentration of 20 LIVI was chosen as the optimum working concentration for the rest of the study.

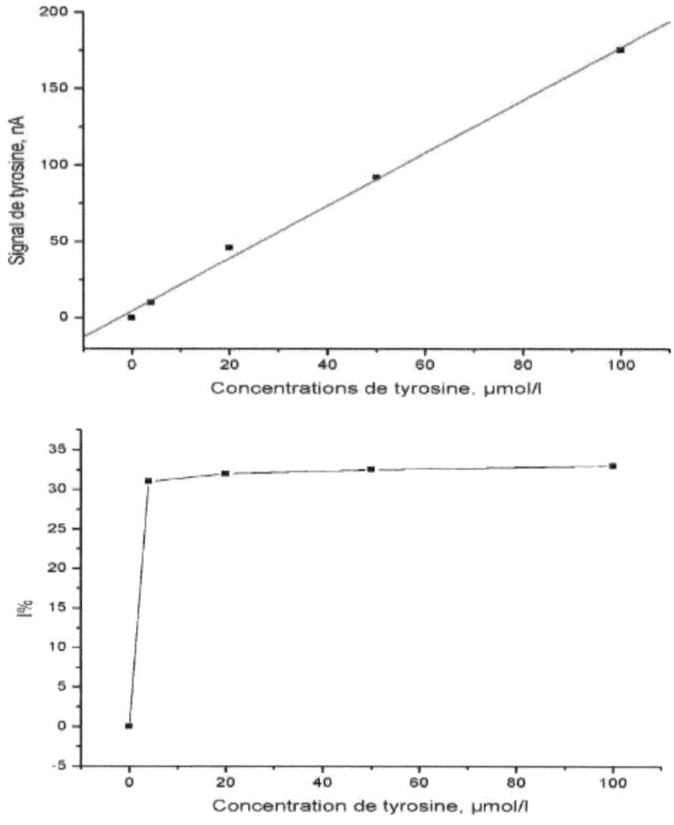

Figure 5: Effect of tyrosine concentration on the (a) Tyrosine signal (b) Percentage decrease in signal in the presence of 1^mol/l of mercury at 0.75 V and pH=7.

III-5 Optimising working conditions

An amperometric study was ëlë carried out in the potential range between 0.6 and 0.9V, the range where a good tyrosine oxidation response had been recorded. Figure 6a shows that a significant decrease of the order of 54% is observed in the presence of 3.0 |imoll-1 of mercury ions at E = +0.75V. It is of the order of 50% at +0.7V and zero at +0.6V.
We also varied the pH of the solution to see if the inhibition rate was pH-dependent. Amperometric measurements were therefore carried out at E=+0.75V and in the pH range 7 to 9. Figure 6b shows that the best result corresponding to 54% tyrosine signal

inhibition is obtained at pH = 7.

We wanted to verify that current inhibition is only possible if tyrosine is oxidised at the electrode surface. To this end, a solution containing 200ццM tyrosine and 30цM mercury (II) was incubated for periods ranging from 10 to 60 min. Injection of 1ml of these solutions into the cell containing 9ml of phosphate buffer pH=7, gives a large mercury oxidation current followed by a decrease in signal that lasts about 11min. This confirms the adsorption of tyrosine followed by complex formation on the electrode surface after the addition of mercury, and suggests that the orientation of tyrosine molecules on the electrode surface is favourable for the formation of a mercury complex adsorbed on the platinum electrode surface.

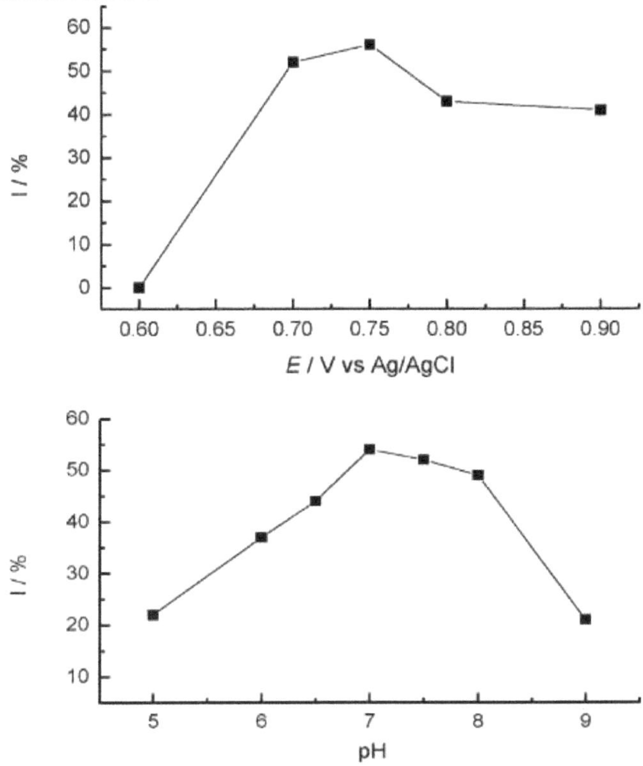

Figure 6: (a) Effect of applied potential at pH= 7 (b) Effect of pH at E=0.75 V vs. Ag/AgCl,

on the percentage decrease in oxidation current of 20 LIVI L- tyrosine + 3LIVI Hg^{2+}

III-6 Calibration curve

Under the optimised conditions prëcëdently (pH=7 and E=0.75V), the variation in the percentage decrease in tyrosine I% signal as a function of the concentration of mercury Hg ions^{2+} ajoutëe, a ële ëtudiëe. The calibration curve shown in Figure 7 shows linearity in the range 0.02 - 3 gmol l^{-1} mercury (II). The regression equation is: log% = 0.38 log $[Hg^{2+}]$ + 3.8 with r = 0.997.

Assuming a background signal-to-noise ratio of 3 (3c), the detection limit is estimated at 14 nmol l^{-1}.

The repeatability of the analysis is determined by performing six repeated measurements with a concentration of 3LIM of mercury. The coefficient of relative standard deviation is 2.2%.

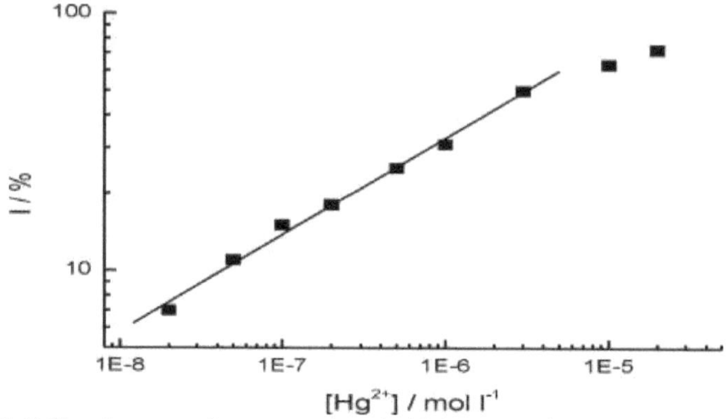

Figure 7: Calibration curve for mercury ions in 20gM L-tyrosine at E=0.75V and pH=7.

III-7 Application to organic mercury

The mëthode optimisëe a ële also testedëe for the dëtermination of traces of the two most responsive forms of organic mercury, methylmercury and phenylmercury. An ampërometric ëtude is rëalisëe under the same conditions as prëcëdently. The addition of a solution of phenylmercury or methylmercury to the tyrosine solution shows no inhibition of the signal. It can therefore be deduced that no complexation is possible with these organic forms of mercury and that this method can only be used for the analysis of mercury ions Hg^{2+} labile, either free or weakly Hëз in complexes.

III-8 Calculation of the complexation constant

The tyrosine oxidation current measured is proportional to the concentration of tyrosine adhered to the surface. When Hg ions^{2+} are added to the solution, they form a complex with tyrosine, and consequently a decrease in current is observed. This decrease is directly related to the concentration of mercury ions and the complex formation constant. We will therefore determine the stcvchionK'tne of the complex and then calculate the complex formation constant.

The formation of the complex between tyrosine and Hg^{2+} can be expressed as:

$$Tyr + n\,Hg^{2+} \overset{\beta_n}{=} [Hg_nTyr]^{2n+} \qquad (1)$$

The training constant, β_n, is

$$\beta_n = [Hg_n\,Tyr]^{2n+} / \{[Tyr].[Hg^{2+}]^n\} \qquad (2)$$

During the expëriences, the total concentration of tyrosine on the Electrode surface, $[Tyr]_T$, has a constant value expressed as follows

$$I_1 = k[Tyr]_T \qquad (3)$$

where k is a constant.

The initial current I_i measured in the absence of mercury is directly proportional to $[Tyr]_T$.

When mercury ions are added to the solution, the current measured is proportional to the concentration of tyrosine, which is not yet complex:

$$I_2 = k[Tyr] \quad (4)$$

What's more, $[Tyr]_T = [Hg_nTyr]^{2n+} + [Tyr]$ (5)

According to equations 2, 3, 4 and 5

$$\beta_n = ([Tyr]_T - [Tyr]) / \{[Tyr].[Hg^{2+}]^n = (I_1 - I_2) / (I_2.[Hg^{2+}]^n) \quad (6)$$

Therefore, the value of n and therefore the strechiomëtrie of the complex can be determined from the slope of the curve of log[$(I_i - I_2)/I_2$] versus log [Hg2+]. A straight line with slope n = 0.51 is obtained.

This value of n indicates that two tyrosine molecules are involved in each Tyr_2Hg electroinactive complex on the electrode surface.

Consequently, Eq.1 can be written as :

$$2Tyr + Hg^{2+} \overset{\beta_{1/2}}{=} [Hg\ Tyr_2]^{2+}$$

Eq.6 can be rearranged as follows:

$$I_1/I_2 = 1 + \beta_n.[Hg^{2+}]^n \quad (7)$$

This allows us to determine the value of $\beta_{1/2}$ from the curve of the variation of I_i/I_2 as a function of the square root of [Hg^{2+}]. The slope of this curve allows us to deduce the formation constant of the mercury-tyrosine complex, $\beta_{1/2}$, which is 627±19.

This value is similar to those obtained for copper and lead complexes with N-(2-hydroxyethyl) piperazine-N'-3-propanesulphonic acid [46].

III-9 Interference study

In order to test the sëlectivitë of this mëthod, the same experiments were carried out under optimised conditions in the presence of various metal cations such as Zn^{2+} , Ag^{+} , Cu^{2+} , Ni^{2+} , Cd^{2+} , Pb^{2+} , Co^{2+} , Fe^{2+} and Fe .$^{3+}$

The concentration of each cation corresponding to a 10% and 50% decrease in current in the presence of 20µM tyrosine was determined.

The results of this study, shown in Table. 1 showed excellent selectivity towards mercury (II) ions.

To obtain a 50% reduction, you need to add a concentration of Zn^{2+} or Cd^{2+} 100 times greater than that of mercury. The only ion that can really interfere with the determination of Hg^{2+} is the copper ion. In fact, according to Table 1, it is sufficient for copper to be present in solution at a concentration 10 times greater than that of mercury ions for a 50% inhibition rate of the oxidation current to be observed. This interference can be avoided by adding a suitable masking agent.

Cation	Hg^{2+}	Zn^{2+}	Ag^{+}	Cu^{2+}	Ni^{2+}	Cd^{2+}	Pb^{2+}	Fe^{2+}	Fe^{3+}	Co^{2+}
$[\ ]_{10}$ (^moll)$^{-1}$	0.05	8.0	6.2	0.7	7.9	5.9	12.3	5.6	10.1	8.2

[ ]$_{50}$ (^moll)$^{-1}$	3	400	330	40	400	330	600	330	500	400

Table 1. Concentration of different cations allowing an inhibition rate of 10%. and 50% of the oxidation current of 20 |imol l^{-1} of tyrosine at +0.75V/Ag/AgCl and pH 7, 0.1M phosphate buffer

III-10 Comparison with other molecules

The behaviour of tyrosine towards mercury ions and its susceptibility to forming a complex was compared with other compounds with a similar structure to tyrosine, such as catechol, dopa, dopac and dopamine; as well as two oxidisable amino acids, cysteine and tryptophan. The structures of these molecules are given in Table 2.

The experiments were carried out at pH 7 and at the potential corresponding to the oxidation potential of each molecule. A cyclic voltammetry study was carried out for each molecule in order to define the working potential. No inhibition reaction was observed for these molecules in the presence of mercury, except for cysteine. In fact, in the presence of mercury, the cysteine response showed a significant and much faster decrease than in the presence of tyrosine.

MOLECULE	STRUCTURE
Tyrosine	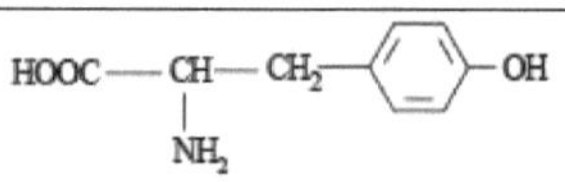
Tryptophan	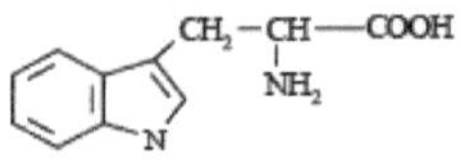
Cysteine	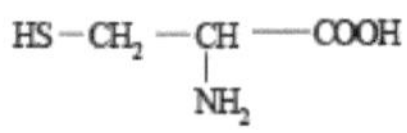
Catechol	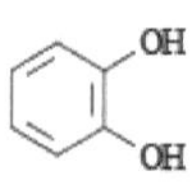
Dopa	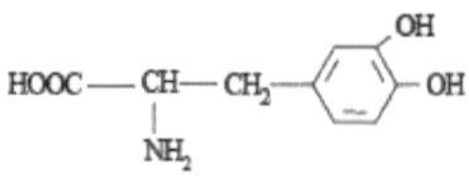
Dopac	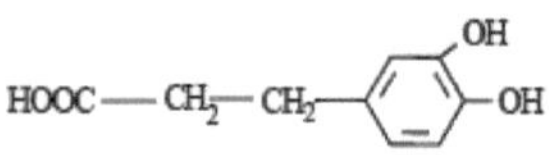
Dopamine	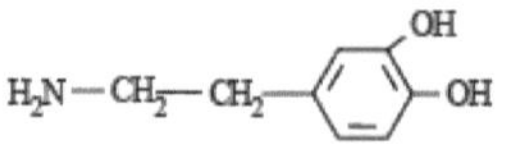

Table 2: Structures of different molecules tested

In fact, cysteine reacts through the thiol groups with mercury in solution to form the mercuric cystëinate complex in solution [47-48], which explains why the response is obtained only 30s after the addition of mercury. This is not the case for tyrosine, where the mercury ions form a complex with the tyrosine oxidised and adsorbed on the surface of the electrode and so the response is not obtained until 11 min after the addition of the mercury ions.

The results, summarised in Table 3, compare the potential for interference by copper ions. They show that cysteine is 3 times more sensitive to mercury ions than tyrosine but, on the other hand, it is much less selective. In fact, a 50% reduction in the cysteine oxidation signal was obtained with a copper concentration 5 times lower than that of mercury.

	Hg^{2+} (µmol l^{-1})	Cu^{2+} (µmol l^{-1})
Tyrosine	3.0	40.0
Cysteine	1.0	0.2

Table 3: The concentration corresponding to the 50% decrease in the oxidation signal of 20 iiM of amino-acid after the addition of metal ions

III-11 Reversibility of complex formation

In order to study the reversibility of the tyrosine-mercury system on the electrode surface, we studied the response of the system after the addition of EDTA. The procedure used was as follows: after recording the tyrosine response, 3LIM of mercury was added to the cell, and a 50% decrease in response was observed. The addition of 20uM EDTA shows a 24% recovery of the initial tyrosine oxidation signal. Addition of a higher concentration of EDTA gave no improvement in signal restoration. This is explained by the formation of the soluble Hg-EDTA complex.

III-12 Analysis of real samples

The analytical performance of the method was first assessed by determining Hg^{2+} in tap water samples. The samples were prepared by adding mercury to tap water and analysed without any prior treatment.

The effect of the addition of each of the three samples on the oxidation signal of 20 LIM tyrosine was studied under the same previously optimised conditions. The results of the determination of Hg^{2+} in these solutions are given in Table 4. The recovery obtained was satisfactory. However, for concentrations below 0.2 LIM, the recovery was lower.

Sample no.	[Hg^{2+}] adds/ (µmol l^{-1})	[Hg^{2+}] found / (µmol l^{-1})	Recovery rate / (%)
1	0,20	0,17	85 ± 3
2	0,30	0,29	97 ± 2
3	0,40	0,42	105 ± 2

Table 4: Analysis of Hg^{2+} in tap water samples

The same method was also applied to three different wastewater samples, where the presence of interferences is obviously possible due to the complexity of the matrix. For this reason, these samples were analysed using the dose addition method. The recovery rates obtained after analysing these samples without any prior treatment were very low (below 40%). The presence of organic matter in the matrix affects the analysis insofar as the mercury ions can be found in their organic form.

Thus the latter are not labile in solution and therefore cannot form complexes with tyrosine. The recovery rate has ële markedly improved after acidifying the ë samples to pH 2. The sample is then IIIërë through a membrane whose pores have a diamëtre of 0.45

|im. The purpose of this pre-treatment is to make the mercury ions labile in the solution. A solution with a concentration of 10 $moll^{-3-1}$ of sodium potassium tartrate is added to the solution as a masking agent[49] to avoid any interference of mëtalic cations with the mercury.
Under these conditions, high recovery rates were obtained (94-102%). The results of the analysis of the various samples to which Hg^{2+} was added are shown in Table 5. Sample 3 was found to contain electroactive organic compounds with an oxidation signal less than 25% that of tyrosine. To calculate the percentage inhibition, the oxidation signal of these electroactive compounds was subtracted from that of tyrosine.

	Sample N°1	Sample N°2	Sample N°3
The concentration detected after dilution of the sample $(nmol\ l\ )^{-1a}$	24 ± 1,5	22 ± 1,3	46 ± 3,4
The concentration detected after the addition of 50nM $(nmol\ l\ )^{-1}$	75 ± 4,5	70.4 ± 4,1	93 ± 8
Recovery rate (%)	102	97	94
Actual value $(^\wedge mol\ l\ )^{-1a}$	0,240 ± 0,015	0,220 ± 0,013	0,460 ± 0,034
Coefficient of variation (%)	6	6	7

[a] the actual values are obtained by multiplying the detected value by the dilution factor

Table 5: Determination of Hg^{2+} in wastewater samples

IV- CONCLUSION

This study led to the development of a new amperometric procedure for the determination of mercury (II). This procedure is based on the formation, on the surface of the platinum electrode, of a complex of formula HgL2 between tyrosine adsorbed and oxidised on the surface, and mercury ions.
This method is very simple and easy to implement. It offers a relatively short analysis time of 11 minutes, is inexpensive and can be applied in the field. An interference study with other metal ions revealed good selectivity of tyrosine towards mercury (II) ions.
A comparative study carried out on other molecules with a similar structure to tyrosine shows that tyrosine is the only molecule capable of forming a complex with mercury ions after being oxidised and adsorbed on the electrode surface. In addition to good selectivity, the detection limit achieved is of the order of 14 nmol $l.^{-1}$
This method can be used with precision for the determination of mercury Hg^{2+} in contaminated water. For very complex matrices, it has been shown to be a reliable procedure for analysing wastewater contaminated with other metal ions or electroactive compounds.
Organic mercury can be analysed by mineralising samples containing organomercury products prior to analysis using the AOAC method [45].

REFERENCES

[1] Y. Bonfil, M. Brand and E. Kirowa-Eisner, Anal. Chim. Acta, 424 (2000) 65.
[2] Yul. D'yachenko, VV. Kondrat'ev, Zh. Anal. Khim, 53 (1998) 401.
[3] HG. Jayaratna, current Separations,16 (1997) 93.
[4] ChM. Watson, DJ. Dwyer, JC. Andle, AE. Bruce, MRM. Bruce, Anal.Chem 71 (1999) 3181.
[5] E.P. Gil and P. Ostapczuk, Anal.Chim.Acta, 293 (1994) 55.
[6] C. Faller, N. Yu. Stojko, G. Henze and K.Z. Brainina, Anal. Chim. Acta, 396 (1999) 195.
[7] I. Svancara, M. Matousek, E. Sikora, K. Schachl, K. Kalcher and K. Vytras, Electroanalysis, 9 (1997) 827.
[8] J. Wang and B. Tian, Anal. Chim. Acta, 274 (1993) 1.
[9] M. Korolczuk, Fresenius J.Anal.Chem, 357 (1997) 389.
[10]E.A. Zakharova, V.M. Pichugina, T.P. Tolmacheva, Zh.Anal.Khim, 51 (1996) 1000.
[11]E.A. Viltchinskaia, L.L. Zeigman, S.G. Morton,Electroanalysis, 7 (1995) 264.
[12]T. Kuwana, W.G. French, Anal.Chem. 36 (1964) 241.
[13]Z. Navratilova, Electroanalysis, 3 (1991) 799.
[14]Z. Navratilova and P. Kula, Electroanalysis, 4 (1992) 683.
[15]E-D. Jeong, M-S. Won, Y-B. Sbim, Electroanalysis, 6 (1994) 887.
[16]S.B. Khoo and Q. Cai, Electroanalysis, 6 (1996) 549.
[17]P. Kula, Z. Navratilova, P. Kulova, M. Kotoucek, Anal.Chim.Acta, 385 (1999) 91.
[18]X. Cai, K. Kalcher, W. Diewald, C. Neuhold and R.J. Magee, Anal. Chem. 345 (1993) 25.
[19]I. Helms, F. Scholz, Fresenius J.Anal.Chem, 356 (1996) 237.
[20]P. Ugo, L.M. Moretto, P. Bertoncello, J. Wang, Electroanalysis, 10 (1998) 1017.
[21]L.M. Moretto, G.A. Mozzocchin and P. Ugo, J. Electroanal. Chem. 427 (1997) 113.
[22]Z. Chen, Z. Pourabedi, D.B. Hibbert, Electroanalysis, 11 (1999) 964.
[23]J-M. Zen, M-J. Chung, Anal Chem, 67 (1995) 3571.
[24]P. Ugo, L. Sperni, L.M. Moretto, Electroanalysis, 9 (1997) 1153.
[25]I. Turyan, D. Mandler, Electroanalysis,6 (1994) 838.
[26]I. Turyan, T. Erichsen, W. Schuchmann and D. Mandler, Electroanalysis, 1 (2001) 79.
[27]T. Marino, N. Russo and M. Toscano, J. Inorg. Biochem, 79 (2000) 179.
[28]R. Andreoli, G. Battistuzzi Gavioli, L. Benedetti, G. Grandi, G. Marcotigiano, L. Menabue and G.C. Pellacani, Inorg. Chim. Acta, 46 (1980) 215.
[29]V. Ryzhov, R.C. Dunbar, B. Cedra and C. Wesdemiotis. J. Am. Soc. Mass Spectrom, 11 (2000) 1037.
[30]D. Van Der Helm and C.E. Tatsch, Acta Cryst, 28 (1972) 2307.
[31]M. Sarat, M. Jesowski, H. Kozlowski, inorg.Chim.Acta, 37(1979)L511.
[32]W.A. Franks, D. Van Der Helm. Acta Crystallogr.Sect.B, 27 (1971) 1299.
[33]A. Alvestrand, M. Ahlberg, P. Forest, J. Bergstrom, *Clin Nephrol*, 19 (1983) 67.
[34]A. J.Gelenberg, C.J. Gibson, *Psychopharmacol Taureau*, 18 (1982)7.
[35]J.S. Meyer, K.M. Agallois, V.D. Deshmuckh, J.AM Geriatr Soc 7(1977)289.
[36]B. Malfoy and J.A. Reynaud, J. Electroanal. Chem, 114 (1980) 213.
[37]G.A. Rivas and V. M. Solis. Anal. Chem. 63 (1991) 2762.

[38]X.-L. Wen, Y. Jia, L. Yang and Z.-L. Liu, Talanta, 53 (2001) 1031.
[39]J. Saurina, S. Hernandez-Cassou, E. Fabregas and S. Alegret. Anal. Chim. Acta, 405 (2000) 153.
[40]K.A. Marx, T. Zhou, J. Electroanal. Chem. 521 (2002) 53.
[41]K. Ogura, M. Kobayashi, M. Nakayama and Y. Miho, J. Electroanal. Chem. 463 (1999) 218.
[42]M. Mikowska, M. Jurikiewicz-Herbich, Pol.J.Chem.72 (1998) 1096.
[43]R.S lojkowska, M. Jurikiewicz-Herbich, Colloids Surf. Part A. Physicochem. Eng.Aspects, 178 (2001) 325.
[44]M. Jurikiewicz-Herbich, A. Muszalska, R. Slojkowska, Colloids Surf. Part A. Physicochem.Eng.Aspect 131 (1998) 315.
[45]Official Methods of Analysis of AOAC, 16th ëdition, Vol II, 1995, Chapter 11, p.17.
[46]T.M. Florence, J.Electroanal.Chem, 97 (1979)219.
[47]Y. Sakane, K. Matsumoko, Y. Osajima, Sci.Bull. Fac. Kyushu Univ. 35(1981) 37.
[48]H.M.V.M. Soares, M.G.R.T. Barros, Electroanalysis 4, 2001, 13.
[49]Jacques TONNEAU, Tables de chimie: un memento pour le laboratoire.

Chapter IV
CHARACTERISATION OF A PATE DE CARBON MODIFIED BY POLY (1,8-DAN) AND ITS APPLICATION TO THE ELECTROCHEMICAL DETECTION OF LEAD

I- INTRODUCTION

In order to meet the growing need for increasingly sensitive and cost-effective electroanalytical techniques and detection methods, the use of modified electrodes, particularly polymers, has certainly been the most surprising development in recent years [1-3].

Nitrogen-containing polymëers, notably polyaniline and polypyrrole, have ëlë widely ëtudiës [4-6]. The important position they occupy is due spëcially to their remarkable ëlectrical, ëlectrochemical and optical properties as well as their multiple applications in different fields. The ëlectrodes modified by these polymëres can be used as sensors, biosensors, catalysts [7-12], etc.

Poly-diaminonaphthatenes (poly-DANs) are polymëers obtained from aromatic monomëers containing two amine groups. Some years ago, Oyama et al [13], Lee et al [14] and Jackowska et al [15] reported the polymerisation of 2,3-DAN, 1,8-DAN and 1,5-DAN. The modification of different solid electrodes by these polymëres was performed by electropolymerisation in acid medium and acetonitrile solution. However, the films resulting from the different polymëers present different properties, due to their different structures.

In 1992, Lee et al showed that poly 1,8-DAN is able to accumulate Ag^{+} ions from solution and subsequently these ions can be exchanged by Hg^{2+} and presumably by other ions, such as Cu^{2+} , Mg^{2+} , Cd^{2+} and Pb^{2+} [14]. Its unique ability to extract heavy metal ions from solution is explained by the fact that one of the two amine groups of the monomëre 1,8-DAN, remains free and probably does not participate in the polymerisation reaction. The chemical structure of the polymer is shown below:

Poly 1,8-DAN

Several mëthods exist for synthesising polymers. However, the most frequently used technique is the oxidation of monomers in order to form a cationic radical followed by an association thus forming the di-cation and the repetition of this phenomenon gives the polymer according to the diagram opposite.

Electrochemical synthesis has rapidly become the general method for preparing polymers because of its reproducibility and simplicity. The advantage of electrochemical polymerisation is that the reactions can be carried out at room temperature. Electropolymerisation is generally carried out in :

- Galvanostatic mode, constant current electrosynthesis;
- Potentiostatic mode at constant potential;
- Scanning or potential scanning.

Palys et al studied the sensitivity of poly 1,8-DAN, prepared on a gold electrode, to Ag^{+} , Hg^{2+} and Cu^{2+} ions using electrochemical and spectroscopic methods[16]. The cyclic voltammetry study shows that Ag^{+} ions are reduced at the polymer surface, while Hg^{2+} or $Hg2^{2+}$ ions are bound to the polymer matrix. Spectroscopic methods (IR, Raman spectra) indicate that the Hg^{2+} and Cu^{2+} ions are complexed by the amine groups in the polymer. The oxidation of 1,8-DAN on gold, silver and copper electrodes was also studied by the same author [17]. This study shows that on the copper electrode we have the formation of the Cu complex^{2+} -(1,8-DAN)n. In order to prove the complexation of Cu^{2+} and VO^{2+} ions by poly 1,8-DAN, Kudelski et al studied the EPR spectra resulting from complexes formed from poly 1,8-DAN, electropolymerised on a platinum electrode, in aqueous solutions of CuSO4 and VOSO4 [18].

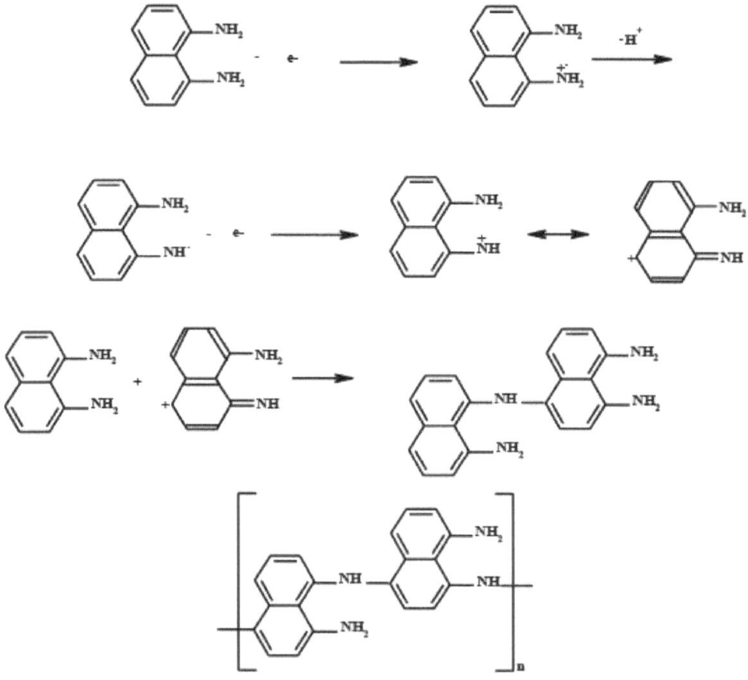

In this work we study the polymërisation of 1,8-DAN within the carbon paste electrode. The modification of carbon paste ëlectrodes with polymëres has ële widely ëtudiëe for analytical purposes. However, polymërisation of a monomëre incorporated within the carbon paste is relatively recent.

Khoo et al were the first to propose this new approach [19]. The carbon paste electrode was ëlë modified with the monomëre 2-methyl-8-hydroxyquinoline. After ëlectropolymërisation, the electrode modified with the insulating polymëre obtained was used for the detection of traces of Cu(II). Such a modification has the advantage of an easily renewable and above all reproducible surface.

In this chapter we prepared poly1,8-DAN within the carbon paste electrode in different acidic media (hydrochloric acid HCl, sulphuric acid H_2SO_4, nitric acid HNO_3, and perchloric acid $HClO_4$).

The polymëre has ëlë compared to meme polymëres synthëtisës on other solid ëlectrodes and on the surface of the unmodified carbon paste lklectrode and where the monomëre was in solution. We also characterised the polymëre by cyclic voltamëtrie and scanning electron microscopy.

Next, we ëtudië the possibility of applying the electrode thus modified by polymëre 1,8-DAN, as a sensor capable of preconcentrating open-circuit Pb(II) ions. The study was carried out using the DPASV technique. Several parameters were studied, such as the percentage of monomer in the paste, the composition and pH of the accumulation medium, the preconcentration time, etc...

II- EXPERIMENTAL PART

II-1 Equipment and reagents

Voltampëromëtric measurements were ëlë rëalisëed using an Autolab potentiostat PGSTAT30 (Ecochemie) controlled by GPES 4.8 software. Platinum, glassy carbon, and modified and unmodified carbon paste ëlectrodes were used as working ëlectrodes. The reference electrode is an Ag/AgCl (3M KCl) electrode. The counter electrode is a platinum wire.

All the chemical reagents used are of analytical quality. Monomdre 1,8DAN was obtained from Sigma. The 10^{-3} mol l Pb (II) master solution^{-1} was prepared from an appropriate quantity of $Pb(NO_3)_2$ (Riedel-de Haen) using bidistilled water.

II-2 Preparation of electrodes

II-2-1 Modification of the carbon paste by the monomer 1,8- DAN: (1,8 DAN-CPE)

The modified carbon paste is prepared by carefully mixing in a mortar 1g of powdered graphite with 1,8-DAN monomer, in appropriate proportions, previously dissolved in a small quantity of acetonitrile. Then 0.3ml of paraffin oil is added to the mixture. The mixture is blended until a homogeneous paste is obtained. The paste is then ready for use once the solvent has evaporated at room temperature.

It should be noted that the homogeneity of the paste plays a very important role in the reproducibility of the results obtained. It is therefore necessary to ensure a homogeneous mixture between the paste and the modifying agent. The paste is then manually inserted into the cavity of the electrode body (L=2mm and ϕ= 3mm), and electrical contact is ensured by means of a steel wire. The surface is finally polished on smooth, clean paper.

II-2-2 Electropolymerisation of 1,8-DAN on the different electrodes:

Two polymerisation methods were recommended in this work. The first by cyclic voltammetry and the second at constant potential.

Poly 1,8-DAN was synthesised in several acidic media on the different electrodes by linear potential scanning (cyclic voltammetry). The potential scanning speed for all voltamograms was set at 50 mV/s.

For potentiostatic polymer synthesis, we applied a constant potential of +0.86V to the electrode for 5min.

III- RESULTS AND DISCUSSION

In order to gain a better understanding of the phenomena observed, the monomer is inserted within the paste. A comparative study was carried out between the polymer formed within the carbon paste electrode and the polymer formed on the surface of the solid electrode.

III-1 Polymerisation of 1,8-DAN on the various electrodes: Monomer in solution

To study the polymerisation of 1,8-DAN, we prepared the polymer on the various solid electrodes and then compared the results obtained.

Poly (1,8-DAN) films were synthesised on the surface of the platinum electrode, the glassy carbon electrode and the bare carbon paste electrode, by scanning the potential between

-0.5 and 1V/Ag/AgCl in a 0.1M HCl solution containing 5mM 1,8-DAN monomer.

The voltamograms recorded during the electrochemical polymerisation of poly1,8-DAN on the different electrodes (fig.1), show oxidation of the monomer at around 0.55V. This irreversible oxidation peak decreases with each scan until it disappears.

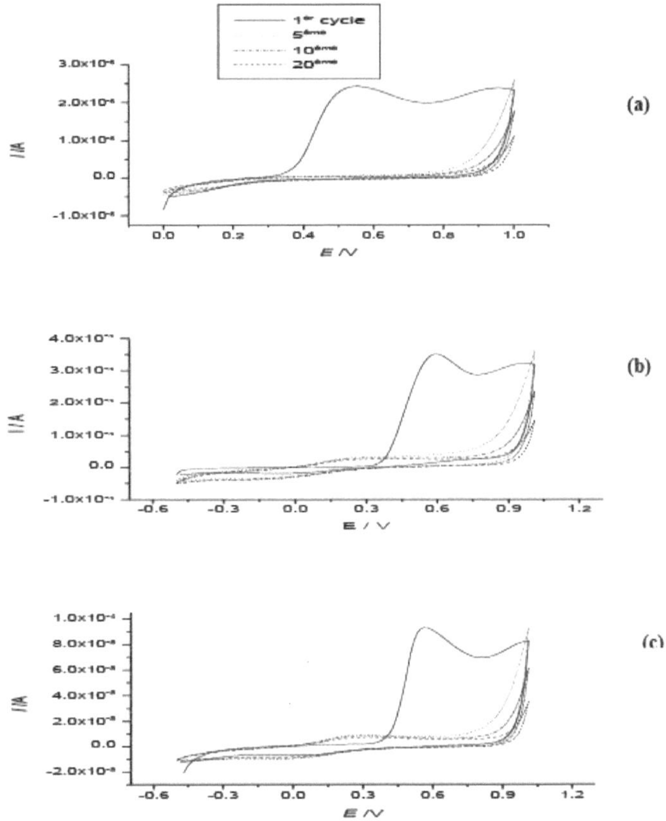

FIGURE 1: Voltamograms obtained in a solution of 5 mM 1,8-DAN at a rate of 50mV/s in 0.1M HCl on the different ëlectrodes. (a) Platinum, (b) glassy carbon and (c) carbon paste electrode.

This indicates that the polymer obtained on each of the ëlectrodes is a non-conductive film which blocks the electrode surface and prevents further polymërisation.

In the presence of 0.1M NaCl in the acid solution containing the monomer, the same result was observed. This shows that the presence of salt in the medium has no effect on the film formed.

In order to prove the formation of the film on the electrode surface, we studied the behaviour of each electrode before and after polymerisation in the presence of the two $Fe(CN)_6$ systems$^{3-/4-}$ and $Ru(NH_3)_6^{3+}$. These two redox systems have been extensively studied in the literature due to the simplicity of the electron transfer properties [20-22].

III-1-1 In the presence of $Fe(CN)6^{3-/4-}$

We recorded the response of the bare electrode and the polymer-modified electrode in a solution of $Fe(CN)_6^{3-/4-}$ 10^{-2} M. The voltamograms obtained for the three bare electrodes are those of a reversible system, although slow on the carbon paste electrode. Figure 2i shows that this response obtained before polymerisation disappears completely when the same experiment was carried out on the modified electrodes, after polymerisation of the 1,8-DAN and formation of the film on the surface. The disappearance of the signal from both systems confirms that the film formed is a non-conductive film that blocks the electrode surface.

III-1-2 In the presence of Ru(NH3)6^{3+}

In order to verify that the phenomenon observed is not a permselectivity effect. We tested the response of the film in the presence of a positive Ru(NH_3)$_6$ system^{3+} . Figure 2ii shows that when the electrode is modified, whatever the nature of the electrode, the 10^{-3} M signal from Ru(NH_3)$_6^{3+}$ disappears completely.

These results are in agreement with those reported by Murphy et al who studied the electropolymerisation of 1.8DAN on the platinum electrode [23].

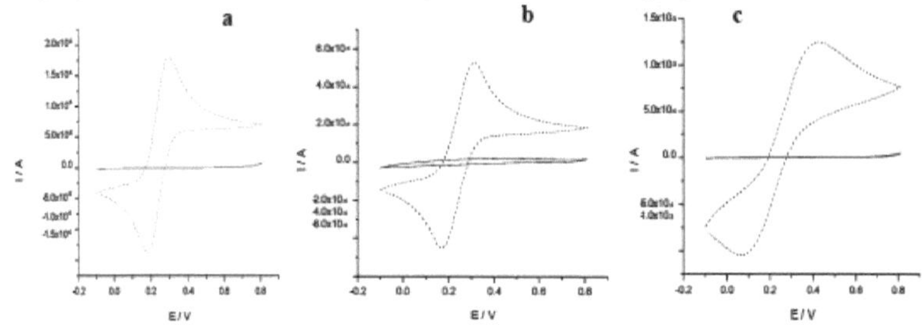

FIGURE 2i: Rëponse of the Ferri/Ferro 10-2M on the bare electrode() and on the ëlectrodes.
modified () by poly 1,8-DAN. **(a)** platinum electrode, (**b**) glassy carbon and (**c**) carbon paste electrode.

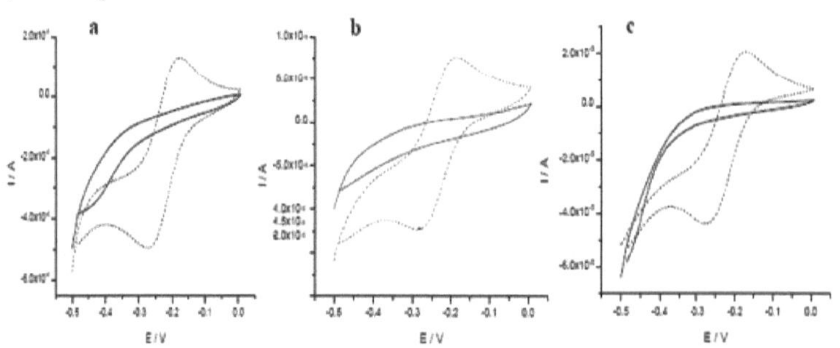

FIGURE 2ii: Response of 10^{-3} M Ru (NH_3)$_6^{3+}$ on the bare (unmodified) electrodes ( ) and on modified electrodes ( ) (a) platinum electrode, (b) glassy carbon and (c) carbon paste electrode.

III-2 Polymerisation of 1,8-DAN monomer incorporated into the carbon paste electrode

After studying the polymërisation of 1,8-DAN on solid ëlectrodes, we studied the polymërisation of 1,8-DAN when it is incorporated into carbon paste, in a proportion of

1%. The study has ël.ë rëalisëe in different media.

III-2-1 Influence of the supporting electrolyte

- In an acidic environment

The voltamograms obtained during the electropolymerisation of 1,8-DAN in hydrochloric acid by scanning between -0.5 and 1V, change radically compared with those obtained previously by polymerising the monomer in solution (figure 3a). The voltamograms show a redox system (E_a= 0.35V and E_c= 0.05V) whose importance increases with each scan. This indicates that the polymer formed is a conductive polymer.

To confirm this electrochemical observation, we studied electropolymerisation in different acid media (nitric, sulphuric and perchloric acid). Figure 3 shows the voltamograms obtained during the electropolymerisation of 1,8-DAN incorporated within the paste in solutions of the different acids. A reversible redox system was obtained by sweeping between -0.5 and 1V, with the peak current increasing with each sweep.

This result was observed for the first time. 1,8-DAN exhibits the same behaviour and conductivity only when the film is electropolymerised in an organic medium of acetonitrile [24]. In our case, it seems that the organic phase of the paste plays the same role as acetonitrile.

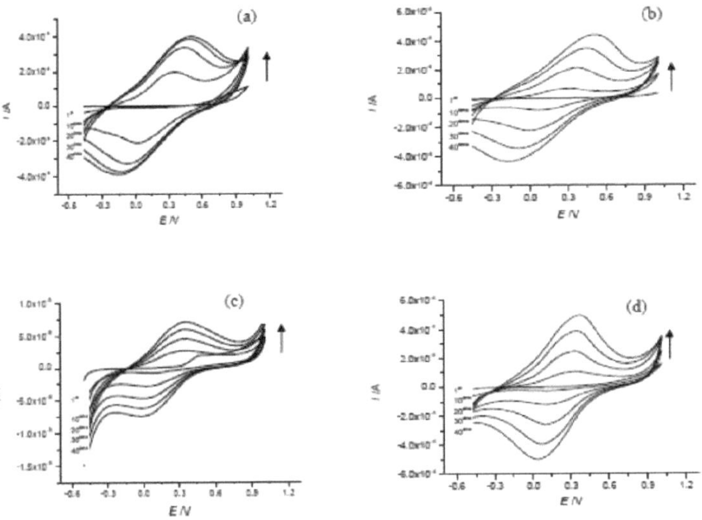

FIGURE 3: ëlectropolymërisation curves for 1,8-DAN incorporated within the EPC electrode in different acidic media (a) 0.1M HCl, (b) 0.1M $HClO_4$, (c) 0.1M H_2SO_4 and (d) 0.1M HNO_3 at a rate of 50mV s .$^{-1}$

In all these solutions, the current increases as the scans progress during Γëlectropolymërisation. Thus, the polymëre continues to form during the first 40 cycles which is ëquivalent to 40 min, then the polymërisation stops and the current remains constant, except for nitric acid where the current continues to increase after the 40min and only stops after 70mn. This behaviour, according to the literature [14], is due to a slow transfer of Electron through the polymëre thickness.

Table 1 shows the peak currents and potentials in the various acids tested. The best ëlectrochemical behaviour was ë1.ë obtained using nitric acid, with an AEp of 300mV . This result is in good agreement with that reported by Lee et al who studied the behaviour of poly-1,8-DAN prepared in acëtonitrile. Voltamograms obtained using an aqueous solution show ëlectrochemically reversible behaviour, with an AEp of 200mV at a scan rate of 50mV/s [14].

Table1: Peak intensities and potentials in different acid media

	Epc (mV)	Ipc (M)	Epa (mV)	Ipa (M)	AEp (mV)	Q (mC)
HCl	-0,156	-391	0,502	399	658	5,74
HNO3	0,038	-496	0,367	497	329	6,00
H2SO4	-0,005	-7,4	0,345	7,1	340	0,16
HCLO4	-0,171	-433	0,517	441	688	5,00

- In a neutral environment

The ëlectropolymërisation of 1,8-DAN incorporates within the carbon paste, a ëlë ëtudië in phosphate buffer. The carbon paste electrode moderated by 1% of the monomëre, a ëlë immërgëe in a solution of 0.1M phosphate buffer at pH 7.4, was scanned for potential between -0.5 and 1V at a rate of 50 mV/s. The Rësults showed that in the phosphate buffer, the film is non-conductive. The voltamograms obtained are broadly similar to those obtained during the electropolymerisation of 1,8-DAN in solution on solid ëlectrodes and the unmodified carbon paste electrode (fig.4).

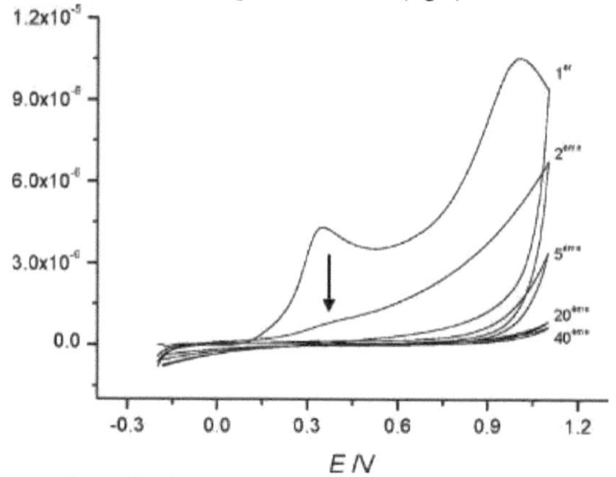

FIGURE 4: ëlectropolymërisation curves for 1% 1,8-DAN monomer incorporated within the EPC electrode in 0.1M phosphate buffer pH=7.4. Scanning speed of 50mV s^{-1}

The same results were found using Tris pH 8, citrate pH 6 and acetate pH 4.7 buffers. This suggests that H+ protons are involved in the redox reaction of poly(1,8-DAN).

III-2-2 Effect of pH

In order to clarify the effect of the concentration of the proton H^+ , and consequently the effect of pH on the polymer formed, we varied the pH between 0 and 2. Figure 5 shows the variation of the anodic peak potential as a function of pH. The Epa was found to vary linearly with pH (r^2 = 0.995) with a slope of 62 mV. This indicates that protons and electrons participate in the polymer redox process in a 1:1 ratio.

By increasing the pH of the solution, the ëlectrochemical activity of poly 1,8-DAN decreases, and disappears readily when the pH of the solution dëpasses 2. However, this activity can be easily recovered when the polymer is put back into a more acidic solution. This result strongly suggests that conduction is proton and electron dependent. An equilibrium that stops at pH>2, beyond pH>4 the film becomes insulating. The same behaviour has been observed for polyaniline [25].

Recorded voltamograms of the polymer in 0.1M HNO3 showed its reversible electrochemical behaviour. The peak currents vary linearly with the square root of the scan rate ranging from 25 to 500mVs^{-1} . This indicates that the redox reaction is controlled by a diffusion process (Fig. 6).

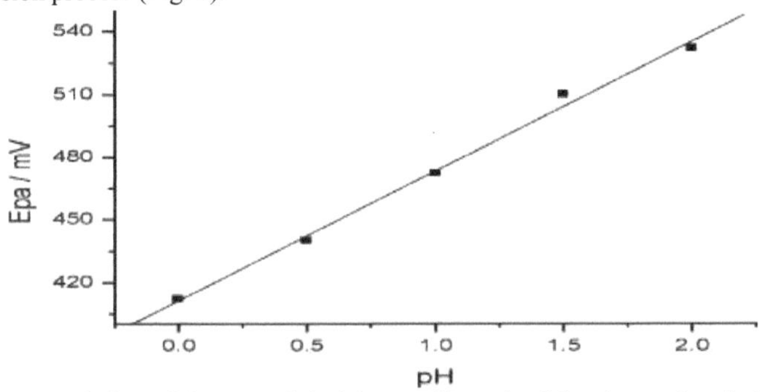

FIGURE 5: Evolution of the potential of the anodic peak of the electrochemical response of the polymer, formed within the carbon paste in HNO3, as a function of pH.

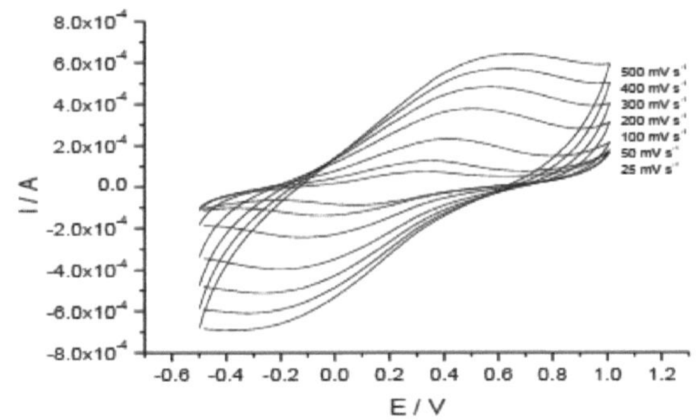

FIGURE 6: The effect of polymer ëlectrochemical response as a function of scan rate in 0.1 M HNO3.

111-2-3 Analysis of the poly1,8-DAN film by scanning electron microscopy (SEM)

We vërifië the morphology of the poly 1,8-DAN film, obtained after ëlectropolymërisation of 1,8DAN incorporated within the carbon paste electrode in nitric acid, by ë scanning electron microscopy (SEM); and then compared it with that of the non-conducting polymer formed after polymerisation of 1,8-DAN in solution on the surface of the unmodified carbon paste electrode.

SEM is used to observe the morphology of electrode surfaces. The principle involves focusing a beam of electrons on a surface and recovering an image from the particles it reflects. Starting with a tungsten filament heated by the Joule effect, incident electrons are produced, known as primary electrons, which are then accelerated. The lenses refine the beam, correct it and focus it on the sample to be characterised. On impact with the sample, the primary electrons create different particles. Secondary electrons from a depth of 10nm provide an image of the surface.

SEM analysis allowed us to see the microstructure of the two polymers formës on the electrode surface. The conductive films of poly 1,8-DAN have a relatively uniform morphology and have a granular structure.

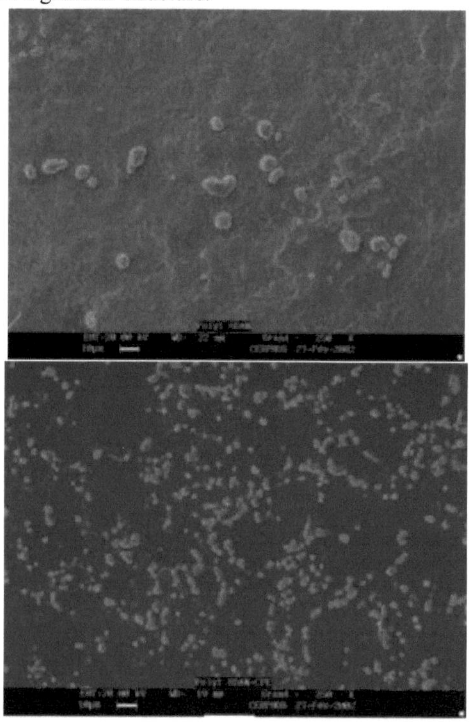

FIGURE 7: Micrograph of poly 1,8-DAN formed from (a) 5 mM of 1,8-DAN in solution on an unmodified carbon paste electrode (b) after electropolymerisation in 0.1M nitric acid of the 1% of 1,8-DAN incorporated into EPC.

111-2-4 Film stability

The stability of poly1,8-DAN, which we synthëtisë within the carbon paste electrode by electropolymerisation in nitric acid, was checked by cyclic voltammetry. The polymer was preserved by leaving the electrodes soaked in phosphate buffer pH=7.4 for 24 hours, after **which** the integrity of the polymer was checked. The voltamograms obtained by scanning the potential on a poly1,8-DAN-modified electrode before and after storage are shown in Figure 8. The latter shows similar voltamograms with a potential shift of 0.1V. This may be explained by an increase in polymer resistance over time.

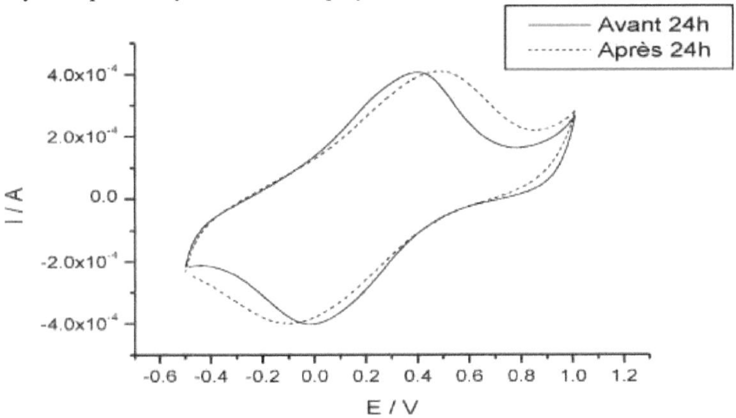

FIGURE 8: voltamograms of poly 1,8 DAN before and after storage for 24 hours in 0.1 M phosphate buffer

III-3 Electrochemical detection of lead using poly-1,8-DAN

Several works have ële carried out on the possibility; of poly 1.8 DAN to form stable complexes with heavy metals [13-14]. In this work we tested the polymer-modified carbon paste electrode for the determination of Pb(II). To optimise the conditions for the electrochemical determination of lead ions, various parameters were studied and optimised.

III-3-1 Procedure for the determination of Pb(II)

We used the carbon paste electrode, modified with poly1,8-DAN, in an aqueous solution of $Pb(NO_3)_2$ containing 0.1M KNO3 at a suitable pH. Preconcentration of the Pb^{2+} was carried out in an open circuit with stirring at 300rpm. After the preconcentration step, the electrode was removed from the solution, rinsed with doubly distilled water and placed in the measuring cell containing a 0.1M HCl solution.

An initial potential of -0.9V was applied for 20s before the voltamogram was recorded using *the anodic differential pulse technique*. The lead peak was observed at -0.56V. The parameters used were a modulation amplitude of 50mV, a modulation time of 50ms, an interval time of 800ms and a potential step of 10mV. No deaeration of the electrolyte solution is required.

The electrode was regenerated by sweeping the electrode between -0.9 and 0 V in a 0.1M HCl electrolyte solution.

III-3-2 Lead accumulation test on different electrodes

Preliminary studies of lead accumulation were carried out on an unmodified carbon paste electrode, a carbon paste electrode modified with the monomer 1,8-DAN without polymerisation and a carbon paste electrode modified with poly1,8- DAN after polymerisation. Figure 9 shows the response obtained by differential pulse voltammetry in 0.1M HCl medium, of the different ëlectrodes testedëes, after open-circuit accumulation of 10-3M Pb^{2+} for 5 min at pH=7. No lead response was found on either the unmodified EPC or the monomer-modified EPC without polymerisation.

However, we have shown that on poly1,8-DAN modified EPC polymerised in two different acid media (HCl and HNO_3), poly1,8-DAN is able to accumulate Pb(II) ions. We observed that these two different films gave a well-defined anodic peak at -0.56V. It therefore appears that the polymer synthesised within the EPC has a high affinity for lead. This result is in agreement with previous work by Lee's team [14]. During electropolymerisation of the 1,8DAN monomer, only one of the two NH2 groups is involved while the other is involved in complex formation.

The best response, in terms of peak intensity, was obtained with the polymer synthesised in nitric acid.

We also tested the sensitivity of the potentiostatically synthesised polymer to Pb ions^{2+} . This polymerisation was carried out by applying a fixed potential of +0.86V to 1,8 DAN-CPE in nitric acid for a period of 5min. The result of the analysis shows an anodic peak obtained at -0.56V after 5min of preconcentration of 10^{-3} M of Pb^{2+} in an open circuit at pH=7. This peak is weaker than that recorded when the film is prepared by cyclic voltammetry.

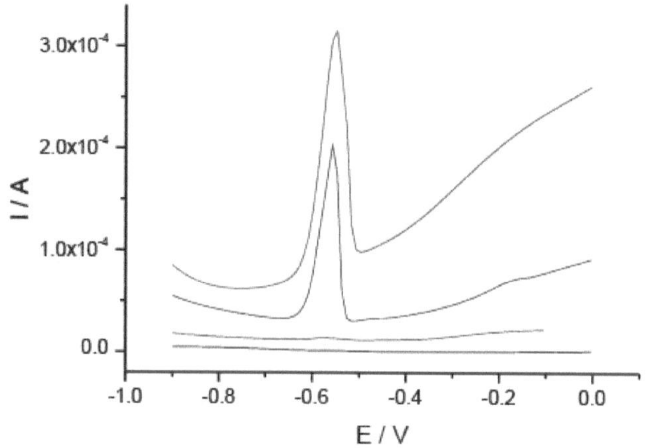

FIGURE 9: The response obtained by differential impulse voltammetry in 0.1M HCl

, on the different electrodes tested, after open-circuit accumulation of 10-3M Pb^{2+} for 5 min at pH=7. On bare EPC (black), EPC modified by monomer (blue), EPC modified with poly 1,8-DAN polymerised in 0.1M HCl (violet) and polymerised in 0.1M HNO3 (red).

III-3-3 Influence of the percentage of monomer in the paste

We have studied the influence of the amount of 1,8-DAN monomer incorporated into the carbon paste on the sensitivity of the modified electrode to Pb ions^{2+} . This study was carried out over a range of monomer percentages from 0 to 3%. The sensitivity of the modified electrodes was evaluated in a 10^{-3} M lead solution after 5min of preconcentration at pH=7. Figure 10 shows the effect of the amount of 1,8-DAN in the electrode on the oxidation current of lead accumulated on the modified electrode by electropolymerisation of the monomer in a 0.1M HNO_3 solution. The results of the study show that the best response was obtained with the electrode modified with 1% monomer.

Indeed, a high current of 300LI A is recorded on an electrode modified with 1% of 1,8 DAN, whereas the current is of the order of 25цA for 3% monomer.

III-3-4 Influence of the pH of the medium

The open circuit preconcentration step is rëalisëe from a 10-3M solution of Pb^{2+} and 0.1M KNO_3. The various pH values of this solution were adjusted with a concentrated solution of HCl and NaOH. The experiments are performed in a pH range varying from pH=3 to 13. Figure 11 shows that the accumulated lead oxidation peak currents increase with the pH of the solution in the pH= 2 to 12 range and begin to decrease thereafter. The best response was obtained after preconcentration in a pH=12 solution. Similar results were reported previously in a study of lead determination using chelating agents such as dithizone [26,27]. Thus 0.1M KNO_3+ 10^{-2} M KOH was chosen as the optimum medium for accumulation. Similar results were obtained using borate and phosphate buffer at the same concentration.

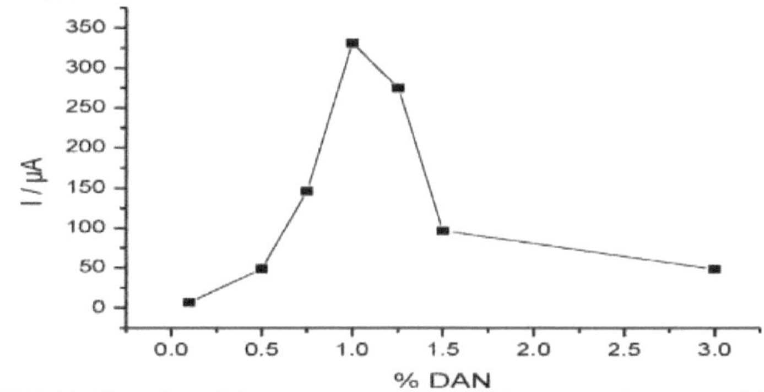

FIGURE 10: The effect of the percentage of 1,8-DAN monomer incorporated into the carbon paste on the intensity of the Pb

carbon paste on the intensity of the 10-3M Pb peak^{2+} after 5 min preconcentration at pH=7.

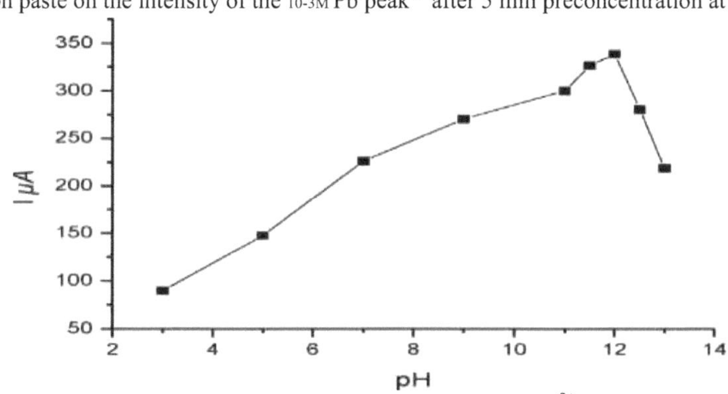

FIGURE 11: Variation in peak current corresponding to 10-3M Pb^{2+} as a function of the pH of the preconcentration solution. Preconcentration time 5min and redissolution in 0.1M HCl.

This result is related to the protonation and dëprotonation of nitrogen groups. This suggests that amine groups, which do not participate in the polymërisation reaction, are able to accumulate lead through a complexation reaction. This behaviour is similar to that proposed by Lee et al [14].

III-3-5 Influence of accumulation time

In order to assess the effect of accumulation time on the sensitivity of the electrode to Pb ions^{2+} , we studied this parameter in two lead solutions at a concentration of 10^{-5} and 10^{-6} moll .$^{-1}$

According to Figure 12 and for the 10^{-6} M concentration, the current increases with accumulation time. For the 10^{-5} M concentration, the current curve of the lead oxidation peak increases linearly as a function of the preconcentration time until a period of 10 minutes is reached, at which point the surface is almost saturated. In order not to extend the analysis time, we considered that ten minutes was an appropriate period for the analysis.

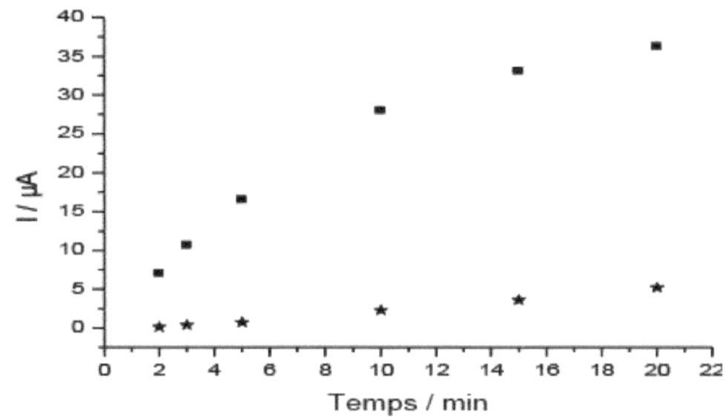

FIGURE 12: Dëdependence of Pb(II) peak currents accumulated on the poly 1,8-DAN-modified electrode obtained in impulse differential voltametry as a function of preconcentration time. ■ 10^{-5} M, ★ 10-6M

III-3-6 Calibration curve

Standard solutions containing Pb ions^{2+} were prepared in 0.1M KNO3 at pH=12. Analysis was carried out using differential pulse voltammetry. After a preconcentration time of 10min, the redissolution step was carried out in 0.1M HCl. Under these conditions, we established the calibration curve by varying the Pb concentrations^{2+} . This curve is shown in Figure 13. The oxidation peak currents increase linearly with low lead concentrations, from 0.2 to 10 |imol l-1 with r2 = 0.998. The detection limit (DL, 3s) is 0.15 |imol l .$^{-1}$

The precision expressed as the relative standard deviation was 6.4% after six repetitions of the measurement of l^mol l^{-1} of Pb^{2+} using the same electrode and 7.8% with 3 different electrodes.

III-3-7 The effect of other heavy metals

Elements that may present some competition for complexation sites on the poly1,8-DAN film were tested using the same procedure optimised for the determination of Pb ions .$^{2+}$

In order to study the effect of the presence of Zn^{2+} ions on the determination of lead, we applied thc same procedure for the determination of ILIM of Pb^{2+} in the presence of Zn^{2+} ions at a concentration of 1ц M. The redissolution peak of Pb^{2+} at E= -0.56V remained constant, but no measurable peak of Zn^{2+} ions appeared up to a high concentration of 5.10^{-5} M (50 times the concentration of lead).

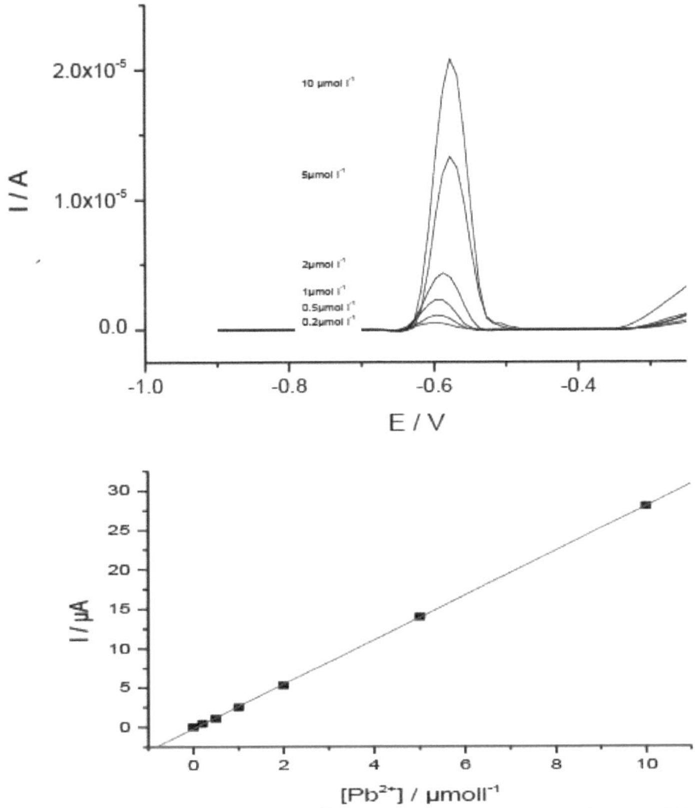

FIGURE 13: Pb ion calibration curve^{2+} after 10 min preconcentration at pH 12 and redissolution in 0.1M HCl.

The study of the effect of Cu^{2+} ions shows that the latter is dëtectaЬle at a concentration above 10-6M. The presence of Cu^{2+} ions gives a well-defined anodic peak at E= -0.46 V, without affecting the lead response. Figure 14 shows the response of a 5.10-5M concentration of copper, resulting in a peak similar in height to that of Pb .$^{2+}$
On the other hand, the presence of cadmium ions, at the same concentration as Pb^{2+} (10^{-6} M), produced no measurable peak but radically affected the response of Pb ions^{2+} . This results in a 52% decrease in the anodic peak of Pb^{2+} . The effect of Cd^{2+} can be ë^Të by masking these ions with sodium oxalate. Thus, for the determination of 1.10'^{6} moll'^{1} of Pb^{2+} , the interference of the same concentration of Cd^{2+} is ëvitëe by adding 10 moll^{-3-1} of sodium oxalate without affecting the lead response.

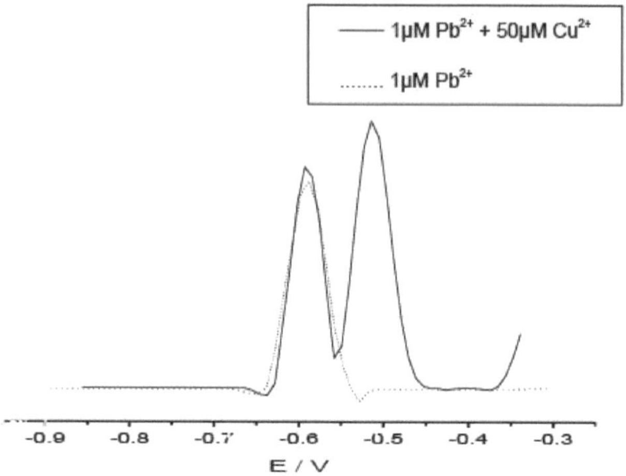

FIGURE 14: The effect of the presence of copper at a concentration of 50^mol l^{-1} on the response of l^mol l^{-1} to lead after 10 min of preconcentration at pH 12.

III-4 Analytical application

The performance of this mëthod has ële subsequently ëvaluatedëe by analysing real ëcHaтШоnз tap water and seawater. We adjustedë these ë samples to pH= 12 and addedë 1mM sodium oxalate to ë eliminate any interference. Seawater was ëlë filtered through a membrane, whose diamëtre is 0.45цт, when a residue was ëlë observed.

The mëthode of dosës additions is usedëe for the dëtermination of Pb^{2+} . For each ë sample five solutions are prepared, 1 ЦМ of Pb^{2+} is addedë to the tap water sample and 1.5цМ to the seawater sample. Tables 2 and 3 summarise the accuracies and recovery rates resulting from the analysis of the five ë tap water samples containing ILIM of Pb^{2+} and the five other ë seawater samples containing 1.5цМ of Pb .$^{2+}$

IV- CONCLUSION

In this chapter we have proposed the use of a conductive polymorphically modified carbon paste electrode for the sensitive and selective determination of lead. We have provedë and this for the first time that ëlectropolymërisation of 1,8DAN within the carbon paste in an acidic medium gives a conductive polymëre, whereas it is not conductive on other ëlectrodes. Such behaviour has ë!ë only been observed by ëlectropolymërising the monomëre in the organic phase and particularly in acëtonitrile.

It has also proved to be a novel, rapid and reproducible method of modifying the electrode surface. This modification makes Electrode, under the optimisëes conditions, sëlective and sensitive for the dëtermination of Pb ions .$^{2+}$

Thus, the analytical application of the optimisëdure on real ëamples confirms the performance of the analysis.

Table 2: Results of the analysis of five tap water samples

	Added concentration iimol/l	**Concentration found iimol/l**	**Recovery rate**

1	1	0,92±0.07	92±7
2	1	0,88±0.05	88±5
3	1	1,02±0.06	102±6
4	1	0,93±0.08	93±8
5	1	1,07±0.1	107±10

96.4 (average)

7.4 (Dë standard deviation)

7.7 (Coefficient of variation)

Table 3: Results of the analysis of five seawater samples

	Added concentration iimol/l	**Concentration found iimol/l**	**Recovery rate**
1	1,5	1,61±0.11	107,3±7
2	1,5	1,53±0.12	102±8
3	1,5	1,42±0.09	95±6
4	1,5	1,37±0.1	91±7
5	1,5	1,32±0.08	88±6

96.6(Medium)

6.9 (Dë standard deviation)

7.1(Coefficient of variation)

REFERENCES:

[1] H. Mark, H. Zhang, A. Galal, J.F. Rubinson, I.H. Ridgway, S.K. Lunsford and H. Zimmer. Electrochimica Acta, 23(1998)3511.

[2] A. Witkowski, A. Brajter-Toth, Anal.Chem.64 (1992) 635.

[3] C.M.A. Brett, D.A. Fungaro, J.M. Morgado, M.H. Gil, J. Electroanal. Chem. 468 (1999) 26.

[4] C. Thiemann, C.M.A. Brett, Synthetic metals 123 (2001) 1.

[5] P. Chandrasekhar, R.W. Gumbs, J. Electrochem.Soc,138 (1991)1337.

[6] M.A. Goyette, M. Leclerc, J. Electroanal Chem, 382 (1995) 17.

[7] S.B. Khoo, R. Ye, Analytica Chimica Acta 453 (2002) 209.

[8] A. Chaubey, K.K. Pande, V.S. Singh, B.D. Malhotra, Anal. Chim. Acta 407 (2000) 97.

[9] J. Yano, K. Ogura, A. Kitani, K. Sasaki, Synth. Met. 52 (1992) 21.

[10]O.A. Sadik, Electroanalysis,11 (1999) 839.

[11]A. Malinauskas, Synthetic Metals 107 (1999) 75.

[12]M. Gerard, A. Chaubey, B.D. Malhotra, Biosensors & bioelectronics 17 (2002) 345.

[13]N. Oyama, M. Sato, T. Ohsaka, Synthetic Metals, 29 (1989) E501.

[14]J. W. Lee, D.S. Park, Y.B. Shim, S.M. Park, J. Electrochem. Soc. 139 (1992) 3507.

[15]K Jackowska, M. Skompska, E. Przyluska, J. Electroanal. Chem 418 (1996) 35.

[16]B.J. Palys, M. Skompska, K. Jackowska, J. Electroanal. Chem. 433 (1997) 41.

[17]B.J. Palys, A. Bukowska, K. Jackowska, J. Electroanal. Chem. 428 (1997) 19.

[18]Kudelski, J. Bukowska, K. Jackowska, J. Molecular Structure, 482-483 (1999) 291.

[19]S.B.Khoo, S.X.Guo, J.Electroanal.Chem, 465 (1999) 102.

[20]C.A. McDermott, K. Keten and Richard L. McCreery, J. Electrochem. Soc. 140 (1993) 2593.

[21]K. Kneten and R.L. McCreery, Anal. Chem, 61(1992) 2518.

[22]N. Motta and A.R. Guadalupe; Anal. Chem. 66 (1994) 566.

[23]L. Murphy, J. Anal. Chem. 70 (1998) 2928.

[24]K. Jackowska, J. Bukowska, M. Jamkowski, J. Electroanal. Chem 388 (1995) 101.

[25]T. Ohsaka, Y. Ohnuki, N. Oyama, G. Katagiri, K. Kamisako, J. Electroanal.Chem, 161 (1984) 399.

[26]T. Molina-Holgado, J.M. Pinilla-Macias, L. Hernandez-Hernandez, Anal. Chim. Acta 309(1995)117.

[27]K.C. Honeychurch, J.P. Hart, D. C. Cowell, Electroanalysis, 12(2000)171.

Chapter V

MODIFICATION OF CARBON ELECTRODES SCREEN-PRINTED WITH A POLYMER MERCURY FILM FOR ANALYSIS OF HEAVY METALS

I- INTRODUCTION

Screen printed electrode (SPE) technology is particularly attractive for the mass production of single-use sensors. These electrodes, characterised by high reproducibility, have been widely used for environmental, biomedical and industrial applications [1-7].

The serigraphy technique

The large-scale production of low-cost, single-use carbon-based electrochemical sensors has therefore been a priority in recent years. Most single-use electrodes are now produced using thin-film deposition techniques: sputtering [8], rotogravure,[9] lithography [10] and screen printing. These techniques involve depositing one or more alternating layers of conductive or insulating materials on a substrate.

Derived from microelectronics, the screen printing of electrochemical sensors has developed rapidly since the early 90s [11]. These electrodes are obtained by depositing a thin layer of conductive ink on an insulating substrate, with the ink passing through a stencil. Its simplicity, high production capacity and the possibility of producing sensors of any size or shape make the screen printing technique one of the most promising technologies [12].

Serigraphy is a technique that involves just one step: passing ink through a stencil to print a substrate placed directly beneath it. Depending on the size of the stencil, a greater or lesser quantity of ink is applied to the resin film.

Using a squeegee, the ink is forced to penetrate the spaces between the free resin threads. The stencil is then lowered onto the substrate to be printed. A second pass with the squeegee enables the chosen substrate to be printed. The stencil is straightened and a new cycle can begin. The printed substrate is dried at room temperature or in an oven to allow the ink to dry. At the end of the operations, the stencil is washed with an ink solvent so that it can be used again later.

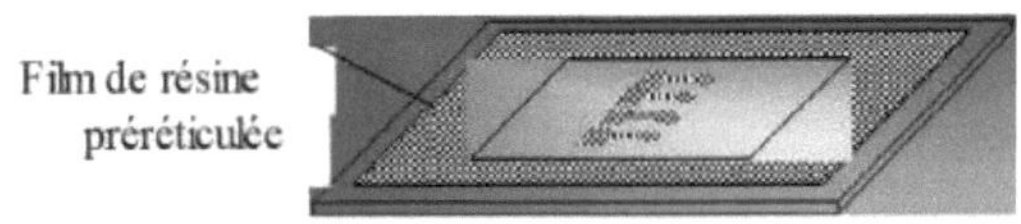

Le pochoir

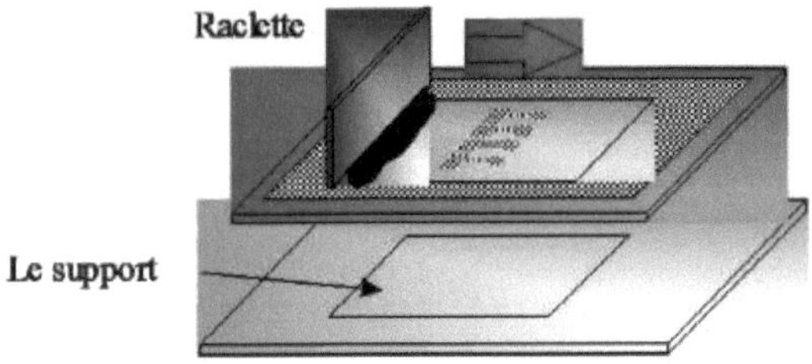

Figure 1: Principle of screen printing: ink is forced through the stencil by the squeegee , printing the substrate.

Manufacture of screen-printed electrodes

The inks were applied to the polyester substrate in the form of films of controlled pattern and thickness. After each film, the substrate was oven dried at 120°C. First the silver tracks are printed. Then graphite is placed on part of the silver track to obtain the working electrode. Finally, the insulating layer, with openings that allow electrical contact with the circuit on one side and the solution on the other, is removed.

The system consists of a graphite electrode (3 mm in diameter), a reference electrode and a silver counter electrode. The cell is printed on an insulating polyester support (450 micron thick). The cell is then cut to size, each measuring 0.8cm x 4.5cm.

It should be noted that the silver reference electrode is more stable when chlorides are present in the solution. Therefore the use of at least 10 mM chloride is suggested and should be maintained constant throughout the analysis. The silver electrical contacts can be covered with a layer of graphite to prevent oxidation. A standard connector is used with these cells.

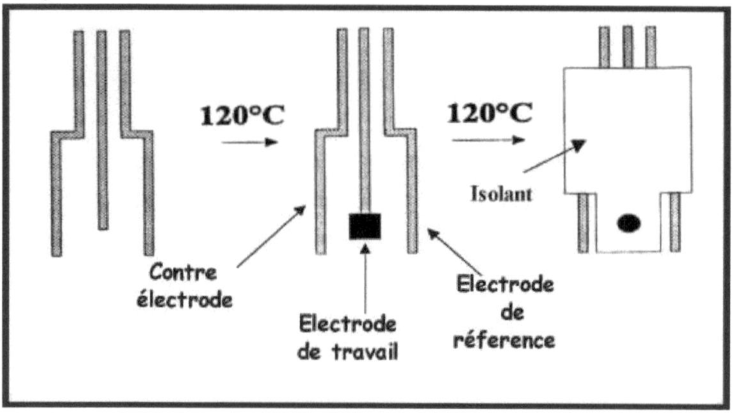

Figure 2: Stages in the manufacture of a screen-printed electrode from carbon

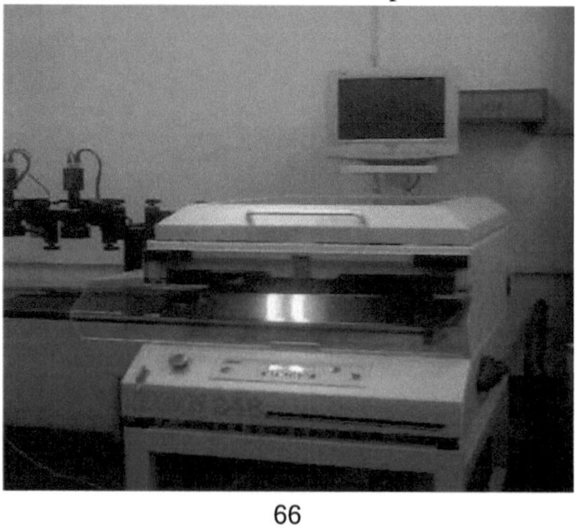

Figure.3 : The machine used to manufacture screen-printed electrodes (**DEK 248**)

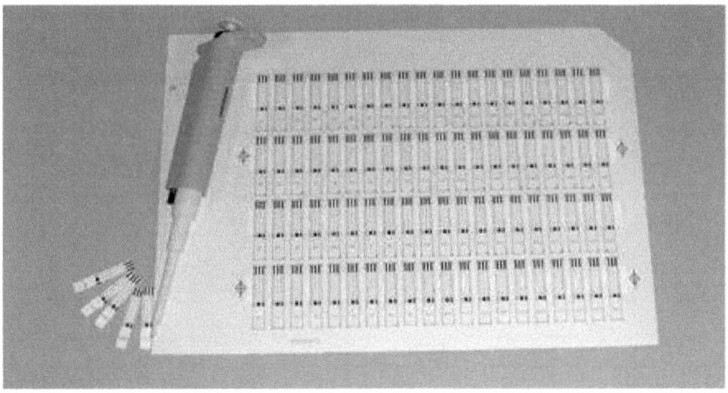

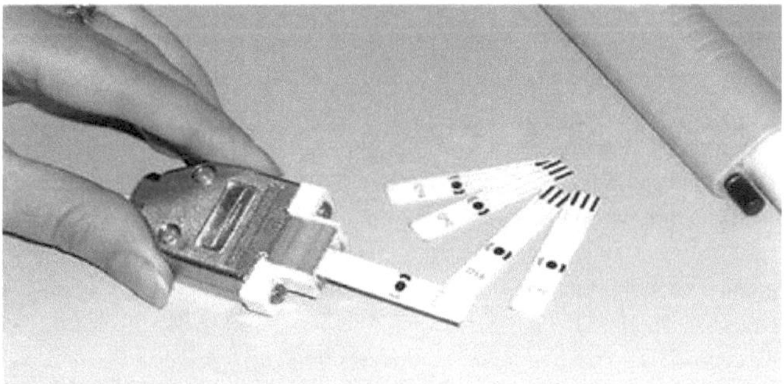

Figure.4: The screen-printed electrodes produced in the laboratory at the University of Florence and the connector used.

Modification of mercury-film screen-printed electrodes for heavy metal analysis

The screen-printed carbon electrodes were also used as a substrate for the mercury thin film, making them very easy to use. Lead, cadmium, copper and other heavy metals that form an amalgam with mercury were therefore detected at trace levels, in a matter of minutes, thanks to the combination of these sensors with highly sensitive electrochemical techniques, such as square-wave anodic redissolution voltametry (SWASV) and redissolution potentiometry (PSA) [13-22].

Electrochemical detection in these analyses is based on voltammetry by anodic redissolution, following the prior deposition of a film of mercury on the surface of carbon screen-printed electrodes using a solution containing a mercury salt. Mercury electrodes in general are often the most widely used electrodes for any electroanalytical analysis. They allow a clean, reproducible surface to be obtained easily, and also allow great access to a wide negative potential range, which is the region where the electrochemistry of most metals takes place. Mercury film electrodes (MFEs) are prepared by the electrochemical deposition of mercuric (II) ions from a solution onto the surface of the carbon electrode.

$$Hg\,(II) + 2\,e \longrightarrow Hg_{electrode}$$

MFEs are often combined with the anodic redissolution voltammetry (ASV) technique. The ions of the metal to be analysed, present in trace form in the solution, are accumulated in the mercury film by electrochemical reduction of the metal atom. The metal atoms then dissolve in the mercury to form an amalgam.

$$M\,(II) + 2\,e + MFE \longrightarrow M(Hg)_{MFE}$$

The oxidation of the metal to be analysed is obtained by applying a potential sweep towards less negative values and so the amalgamated metal redissolves when the potential corresponds to its reoxidation. The intensity of the resulting current depends on the concentration of the metal atom in the MFE, which in turn depends on the concentration of the metal ion in the sample. The potential at which each metal ion is oxidised allows the metal to be identified. Quantitative and qualitative analysis is therefore easy.

Oxidation leads to a redissolution or rather a return of the amalgam metal ions on the mercury film to the solution.

$$M\ (Hg) \longrightarrow M\ (II)_{soiution} + 2\ e + MFE$$

However, the use of a toxic mercury solution, necessary for the formation of the film on the surface of the electrode in the field, presents a major environmental problem. With this in mind, and with the aim of developing new electrochemical sensors for the simultaneous analysis of several amalgamable trace elements, prepared in the laboratory and ready for use in the field, we have used screen-printed carbon electrodes modified with a mercury acetate salt and protected by a polymer.

Currently, various polymers with interesting properties are commercially available and are widely used in the development of electrochemical sensors and biosensors. They are chemically inert, non-electroactive, hydrophilic, and some are even insoluble in water, making them very useful for protecting the electrode surface [23-26].

In order to address the environmental problems associated with mercury contamination in nature, the performance of various polymers used to protect electrodes for heavy metal analysis was analysed. A comparative study was carried out between Nafion®, Eastman Kodak AQ29® (perfluorosulphonate polymers) and Methocel® (a cellulose derivative).

Such modifications have already been reported in the literature. To modify the electrode surface, Brainina et al proposed a soluble mercury compound ($Hg(Ac)_2$) protected with a Nafion® layer, as well as the use of insoluble organic mercury complexes, for the determination of Cd, Pb and Cu [27,28]. Faller et al used these same electrodes for the analysis of tin (IV) [29]. A satisfactory correlation between the added and found concentration of metal ions was obtained over a long period of time, making these electrodes promising for the analysis of heavy metals. In addition, it has been shown that these modified electrodes are much less polluting than ordinary graphite electrodes conventionally modified with a thin film of mercury [2729]. Dalagin [30] and Macca [31] have also reported the use of a Nafion - Mercury Acetate membrane on a vitreous carbon electrode.

Screen-printed carbon electrodes modified by a mercury acetate salt trapped in membranes were used in this work. The techniques used to determine trace metals were square-wave anodic redissolution voltammetry and redissolution potentiometry (PSA). These two techniques have the advantage of being very little affected by oxygen, so no aeration step is necessary, in line with the requirements of analyses intended for field applications.

For SWASV, the square wave shape facilitates redissolution analysis in the presence of dissolved oxygen [32]. This is due to the depletion of oxygen at the surface prior to rapid scanning. Oxygen reduction takes place during metal reduction at the electrode surface, by applying a sufficiently negative potential during deposition. If the scan is completed before a significant amount of oxygen diffuses to the electrode surface, the resulting voltammograms should be similar to those obtained from an aerated solution [33]. Therefore the analysis time can be reduced by avoiding any deaeration of the medium.

PSA has the same advantages as the above technique, sample deaeration is generally unnecessary as the dissolved oxygen can be used as an oxidant. Oxidation (carried out in this work at a constant current of I to A) redissolves the amalgamated metals on the electrode; and the metals diffuse in their ionic form into the solution. Measurement of the electrode potential as a function of time provides both quantitative and qualitative information on the metals present in the solution[34].

II- EXPERIMENTAL PART

II-1 Equipment

Square-wave anodic redissolution voltammetry (SWASV) and redissolution potentiometry (PSA) were carried out using the Autolab PSTAT 10 system (Ecochimie, Netherland).

Carbon screen-printed electrodes, with a silver reference electrode, a graphite counter electrode and a graphite working electrode, are prepared using a DEK Model 245 screen-printer (Weymouth, England), using various inks from Acheson Italiana (Milan, Italy). These electrodes are obtained by depositing alternating layers of graphite ink (Elettrodag 423), silver ink (Elettrodag 477 ss rfu) and an insulating ink (Elettrodag 6018 ss) on an insulating polyester support [35].

II-2 Reagents

Nafion® 5% w/v solution in a low aliphatic alcohol/water solution was supplied by Aldrich; Eastman-Kodak AQ29D was supplied by Eastman-Kodak (USA); Methocel® 90HG was purchased from Fluka. Mercury acetate, mercury chloride, 96% ethanol, Suprapur hydrochloric, nitric, acetic and perchloric acids are all from Merck. The water used to prepare the solutions was demineralised water purified by passing through a Milli-Q filtration system (Millipore, Italy). The heavy metal solutions (Cd^{2+} and Pb^{2+}) were prepared by diluting standard solutions of atomic absorption spectroscopy quality (Fluka).

II-3 Electrode modification

a- Coating of the serigraphed carbon electrode with Hg(II) and Methocel® HG90 **:**

The carbon screen electrode was modified using a solution containing 100mg mercury acetate and 100lil acetic acid in 10ml H_2O (solution A). To 1.5 ml of this solution, 3.5 ml of H_2O and 125 mg of Methocel® HG90 were added. Using a propette, 5 l of the solution is applied to the surface of the working electrode. The film is air-dried at room temperature.

b- Coating of the serigraphed carbon electrode with Hg(II) and Nafion® :

To 1.5ml of solution A is added 3.5ml of Nafion prepared in 10ml of 96% ethanol. 5LI1 of this solution is deposited on the surface of the working electrode. The film is air-dried at room temperature.

c- ***Coating of the serigraphed carbon electrode with Hg(II) and Kodak®AQ29D*** :

To 1.5ml of solution A is added 3.5ml of Kodak®AQ29D prepared in 10ml of 96% ethanol. 5LI1 of this solution is deposited on the surface of the working electrode. The film is air-dried at room temperature.

d- Covering the carbon screen-printed electrode with a thin film of mercury

The thin film of mercury was formed on the surface of the working electrode from a solution of 100ppm mercury chloride in hydrochloric acid with stirring, by applying a potential of -1.1V for 2 min. A potential of -0.2V was then applied for 2 min to clean the electrode surface.

II-4 Analytical procedure

Square-wave anodic redissolution voltammetry (SWASV) and redissolution potentiometry (PSA) were carried out using the same parameters optimised by Palchetti et al [10].

For SWASV, the conditioning step is carried out at -0.3V for 60s, followed by metal deposition at -1.1V for 120s and an equilibrium time of 30s. The square wave has an amplitude of 28mV, a step size of 3mV and a frequency of 15Hz.

For PSA , Conditioning is at -0.3V for 60s, deposition is carried out at - 1.1V for 120s, an equilibrium time of 30s and a redissolution current of +1μA.

Each modified electrode is conditioned before use by applying a potential of -1.1V for 300 s to the electrode surface in a 0.1M hydrochloric acid solution and then performing 10 voltammetric scans until a low background is obtained. Figure 5 shows the background before and after conditioning.

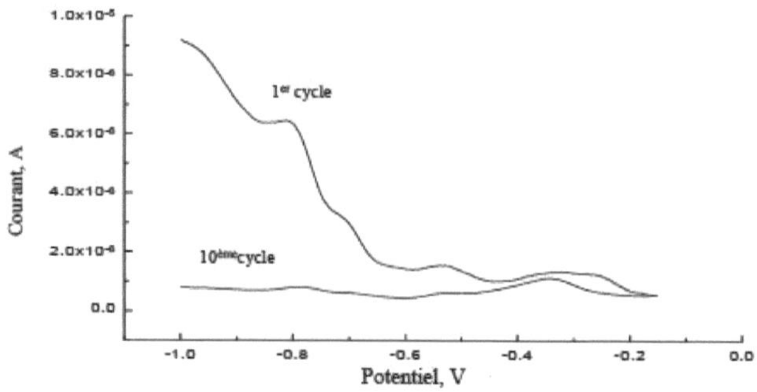

Figure.5: Conditioning of the modified electrode after 10 cycles in a solution of hydrochloric acid solution 0.1M

III-RESULTS AND DISCUSSION

The various polymers that have been studied for the development of mercury sensors for the analysis of heavy metals in the field are: Nafion®, EastmanKodak® AQ29D, Methocel®. Each electrode modification was carried out using the same amount of mercury acetate in the polymer solution. The ratio of mercury to polymer in the solution used for all experiments, irrespective of the polymer, was chosen in accordance with data reported in the literature [30], where an optimised value of 30% mercury / 70%polymer solution (v/v) was given.

III-1 Characterisation of the different modified electrodes

For each modified electrode, various parameters were studied: sensitivity, peak resolution and reproducibility in HCl (the most widely used electrolyte support for heavy metal electrochemical analysis). These experiments were carried out using the PSA technique with a redissolution current of +1 LI A. A preliminary study was carried out with different lead standard solutions at different HCl concentrations.

III-1-1 Nafion - mercury

The electrodes modified by the Nafion-mercury membrane were studied in a standard solution of Pb at a concentration of 150ppb in 0.12M HCl medium. Figure 6 shows the potentiogram obtained, with a clear lead peak at -0.5V.

The behaviour of this electrode in an acid medium was studied at different HCl concentrations. The variations in peak area as a function of lead concentration, in HCl solutions ranging from 0.12 to 1.2M, are shown in Figure 7. It can be seen that as the acid concentration increases, the sensitivity of the Nafion-mercury membrane-modified electrode decreases markedly.

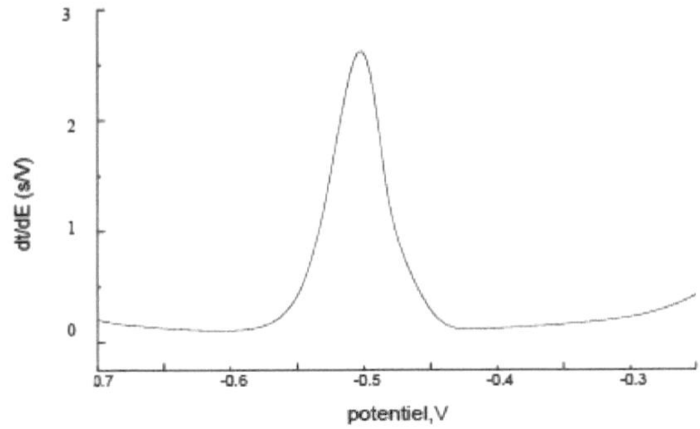

Figure.6: Potentiogram corresponding to 150ppb of Pb(II) from a Nafion-mercury membrane electrode in 0.12M HCl medium. Conditioning: 60s at -0.3 V, deposition: 120s at -1.1V redissolution at +1μA.

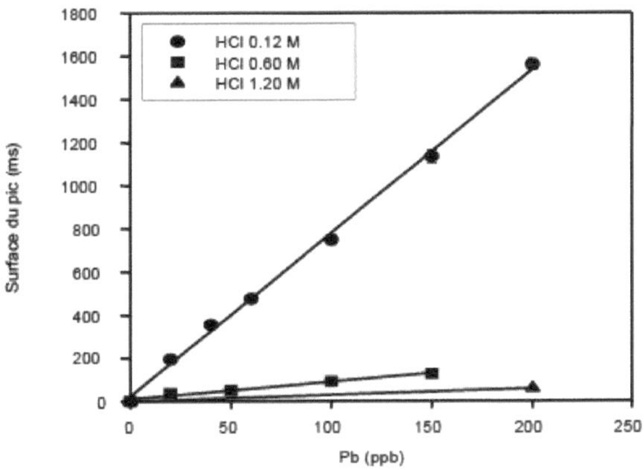

Figure.7 : Lead calibration curves on the membrane-modified electrode Nafion®-mercury at different HCl concentrations: 0.1 M, 0.6 M, 1 M

III-1-2 Eastman Kodak- Mercury

A study of the behaviour of the electrode modified by the Eastman Kodak-Mercury membrane was carried out, in Pb solutions with concentrations of up to 200ppb and at a concentration of 0.1M HCl. Very low sensitivity and poor peak resolution were obtained even at low acid concentrations.

The potentiograms of lead (II) at different concentrations from 0 to 200ppb, in 0.1M HCl, are shown in Figure 8.

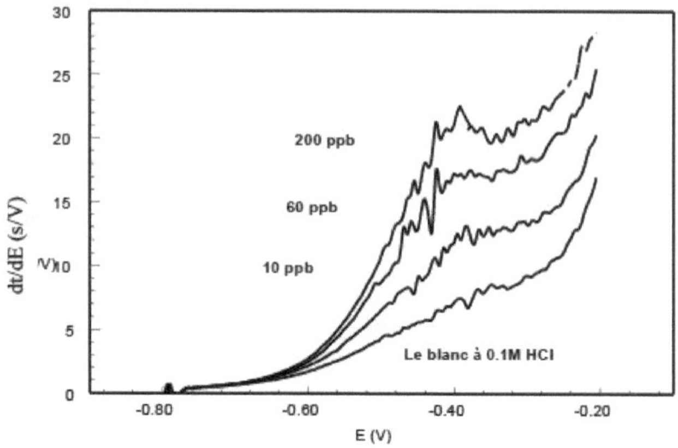

Figure 8: Potentiograms of Eastman-Kodak® AQ 29D-mercury electrode in mercury electrode in 0.1 M HCl medium at different lead concentrations.

The low sensitivity of Nafion- and Eastman-Kodak-modified electrodes in acidic media can be explained by changes in the morphology and geometric parameters of the polymer microstructure with increasing ionic strength. This phenomenon has already been observed and reported in the literature for perfluorosulphonate polymers [20,21]. These changes could strongly influence the behaviour of polymers tested in a concentrated HCl solution for redissolution analysis.

The behaviour of these polymers can also be explained by the shape of the lead complex in solution, depending on the concentration of HCl and the charge of the polymer. Nafion and Eastman Kodak are perfluorosulphonate-based polymers, so they are negatively charged cation exchangers. And generally in solutions

Concentrated with HCl, lead is in the form of a negatively charged complex. Electrostatic repulsion could be the reason for this behaviour.

III-1-3 Methocel - mercury

The sensitivity of electrodes modified with a methocel-mercury membrane to lead was then tested. This electrode appears to be more interesting than the previous two. The potentiograms of Pb(II) obtained in the 0 to 120ppb range, in 0.1M HCl, are shown in Figure 9a. We therefore observe well resolved peaks and high sensitivity.

To check the sensitivity of the electrode modified by a methocel-mercury membrane in a very acidic environment, we compared the response of different lead concentrations in 0.1, 0.6 and 1 M HCl.

The calibration curves for lead (II) in three HCl solutions of different concentrations are shown in Figure 9b. Regardless of the concentration of the acid solution used, the results show no significant difference. The methocel-mercury membrane can therefore be used in HCl-concentrated solutions without any reduction in sensitivity.

So, since the Methocel-mercury membrane can be used over a wide range of chloride concentrations, and since the use of this polymer is not yet reported in the literature in redissolution analysis, we decided to characterise these modified electrodes and test their performance, in standard solutions and in real samples. A comparison with conventional mercury thin film electrodes was also considered.

(a)

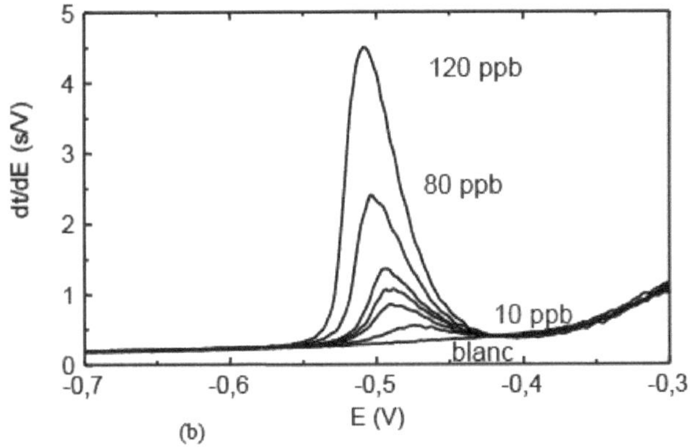

(b)

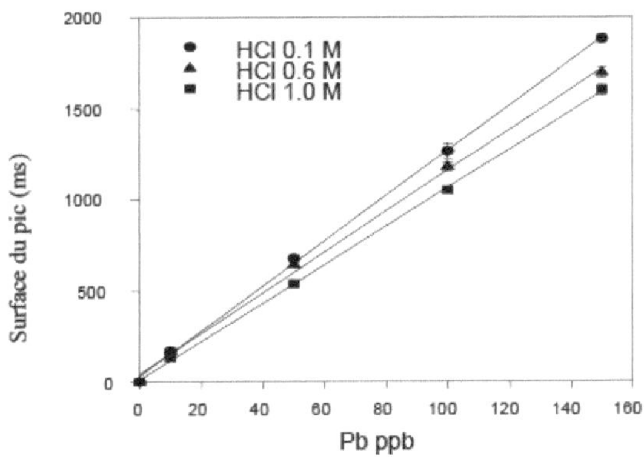

Figure.9: (a) Potentiograms of the methocel-mercury membrane electrode in 0.1M HCl at Pb concentrations^{2+} of 10, 20, 30, 40, 80 and 120 ppb; (b) Calibration curves for the same electrode in acid medium, with concentrations of 0.1 M, 0.6M and 1M HCl.

III-2 Use of Methocel-mercury modified screen-printed electrodes for the simultaneous analysis of lead and cadmium

The analytical performance of this type of electrode was tested in the presence of Pb^{2+} and Cd^{2+} in solution. This study was carried out using square-wave voltametry as the electroanalytical technique.

On this electrode, the two elements give well resolved and well separated peaks (-0.55V and -0.72V for Pb^{2+} and Cd^{2+}). Calibration curves for lead and cadmium were established in 0.1 M HCl. Good sensitivity was observed for both elements as shown in Figure.10.

The detection limit, after two minutes of accumulation under agitation, is 0.8 ppb and 1 ppb respectively for lead and cadmium. The value of the detection limit was calculated as 3 times the noise measured at 10 ppb for lead and cadmium.

The calculated relative standard deviation (R.S.D.%) is 2%, the experiment being repeated 10 times on the same electrode with a 10 ppb lead solution, and an R.S.D.% of 10% with six different electrodes (n=6).

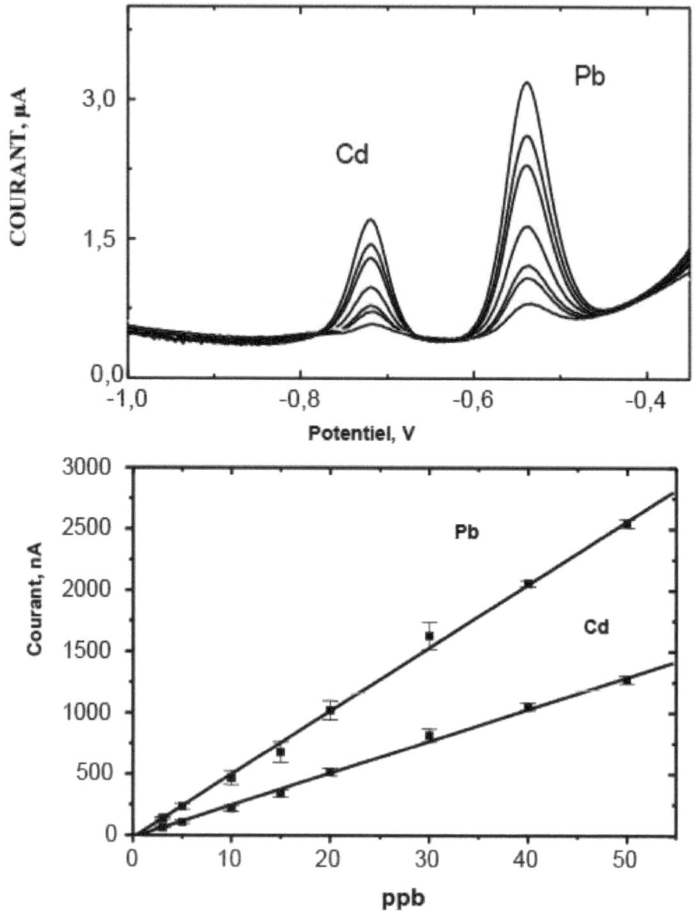

Figure.10: Calibration curves for lead and cadmium on the methocel-mercury membrane electrode in 0.1 M HCl, using the SWASV method

III-3 Electrode stability

We then studied the stability of the electrodes modified with the Methocel-mercury film. The modified electrodes were stored at 4°C in the dark. Accumulation and measurements were then carried out under the same conditions as above. The lead calibration curve obtained with an electrode stored for three months was compared with that obtained with a freshly prepared electrode. Figure 11 shows that the electrode remains stable throughout this period.

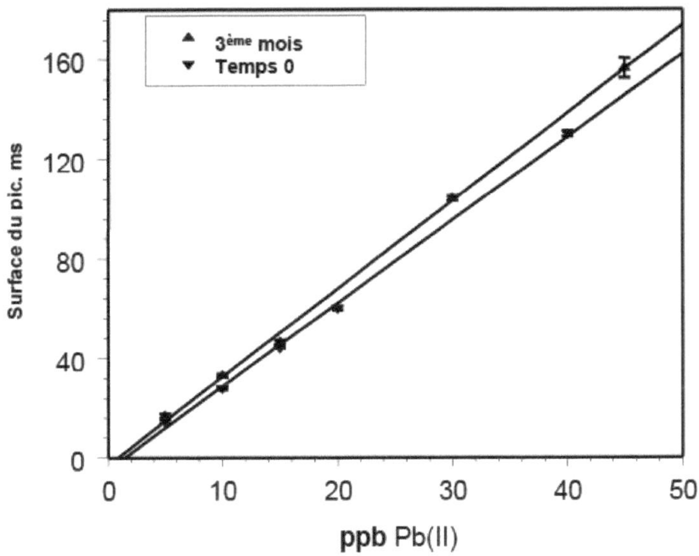

Figure 11: Stability of Methocel®-mercwe membranes: lead calibration curves obtained with the same electrode before and after 3 months. PSA: + 1 μA. Depot of 120 s at -1 V.

III-4 Analysis of real samples

Methocel-mercury membrane-modified carbon screen electrodes have been tested for the analysis of lead and cadmium in three different medicinal plants.

III-4-1 Preparation of samples before analysis

Before analysing the samples, mineralisation by perchloric nitric attack was optimised after several experiments [36]. The mineralisation was carried out in a closed tube. To 0.1 g of sample was added 3.2 ml of nitric acid and 0.8 ml of perchloric acid; the whole was heated to 200°C until a white residue was obtained. 5 ml of the supporting electrolyte (0.1M HCl) was added before starting the analysis.

To avoid any matrix effect, the analysis was carried out using the dose addition method.

III-4-2 Analysis of samples

The analysis of lead and cadmium in the 4 samples was carried out using the dose addition method. Figure.12 shows the result obtained on sample.1 after dilution. Each standard addition consisted of 5ai of an 8mg/l solution of Cd^{2+} and Pb^{2+} . The final concentration in the solution to be analysed was 8ppb. The concentration of lead and cadmium in the original samples was quantified using the standard addition curve.

From this standard addition curve, we can deduce the concentrations of the two elements in the analysis sample: [Cd^{2+}]=4ppb and [Pb^{2+}]=8ppb.

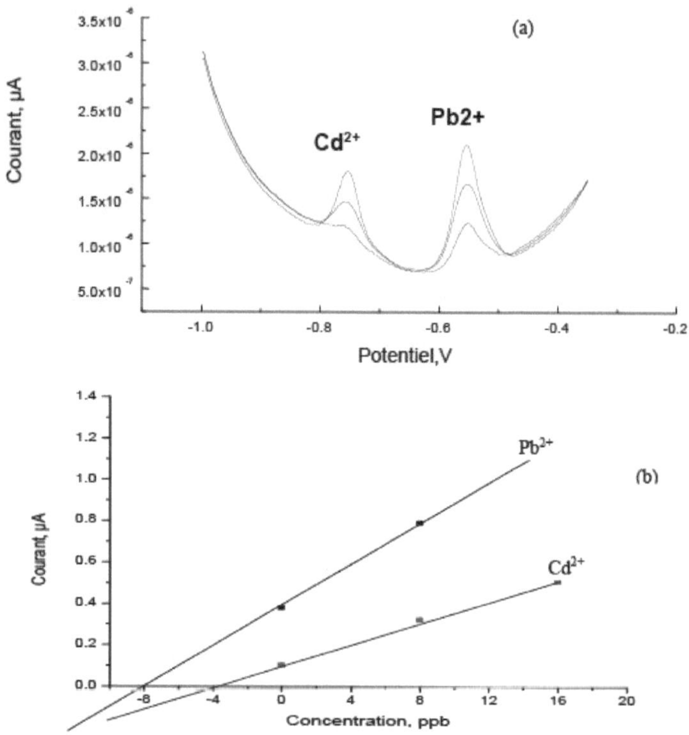

Figure 12: (a) Voltamograms of the lead and cadmium dose additions to one of the samples to be analysed.

samples to be analysed (b) the corresponding calibration curves.

In the same way, we analysed the other three samples. The mean values obtained after 3 measurements of each sample are shown in Table 1 below.

Table 1: Results of the analysis of the 4 samples

Sample No.	$[Cd^{2+}]$ / ppb	$[Pb^{2+}]$ / ppb
1	*4 ± 1*	*7 ± 1*
2	-	*9 ± 3*
3	*13 ± 1*	*123 ± 21*
4	*33 ± 1*	*299 ± 33*

III-5 Comparative study

In order to assess the method's performance, the cadmium and lead values obtained after analysing the various samples were compared with those obtained using mercury thin film modified (MTFE) screen-printed electrodes.

The modified electrodes were prepared as described above. The samples were treated in the same way, i.e. mineralisation by nitric-perchloric acid attack. The results obtained after analysis of the four samples are shown and compared with those obtained by analysis with the screen-printed electrodes modified with methocel-mercury (MME) film, in Table 2.

A comparison between the values obtained by the two methods shows that there is a very good correlation.

Table 2: Results obtained using a mercury thin film electrode and an electrode modified by methocel-mercury film

	MTFE		MME	
Ech. N°	Cd (PPb)	Pb(PPb)	Cd(PPb)	Pb(ppb)
1	*4 ± 1*	*8 ± 1*	*4 ± 1*	*7 ± 1*
2	-	*11 ± 3*	-	*9 ± 3*
3	*16 ± 2*	*113 ± 3*	*13 ± 1*	*123 ± 21*
4	*30 ± 1*	*271 ± 39*	*33 ± 1*	*299 ± 33*

IV- CONCLUSION

In this work we have shown that Methocel, a cellulose derivative, could be an interesting alternative for the development of mercury thin film modified electrodes for the detection of heavy metals. Methocel-protected mercury-modified electrodes, compared with other membranes studied in this work, are more suitable for use in very acidic environments.

The use of such a modification instead of a conventional mercury thin film meets international requirements and avoids the need to use the mercury solution during field analysis. In fact, these membranes are pre-deposited on the surface of the electrode in the laboratory and ready for any decentralised application. In addition, this procedure requires a much smaller quantity of mercury than that used to form the thin film of mercury. A few microlitres of solution are sufficient to cover the active surface of the electrode modified with a methocel-mercury film, instead of a few millilitres in the case of the conventional thin film.

REFERENCES

[1] J-Y. Choi , K. Seo , S-R. Chob, J-R. Oh,S-H. Kahng, J. Park . Anal. Chimi. Acta 443 (2001) 241.

[2] S. Cosnier, C. Gondran, R.Wessel, F-P. Montforts, M. Wedel, J. Electroanal Chem 488 (2000) 83.

[3] A. Avramescu , T. Noguer , V. Magearu , J-L. Marty, Anal. Chim. Acta 433 (2001) 81.

[4] J-M. Zen, C-C. Yang, A. Senthil Kumar. Anal. Chim. Acta 464 (2002) 229.

[5] L. Authier, C. Grossiord, P. Brossier, B. Limoges .Anal. Chem,73(2001)4450.

[6] O. Bagel, B. Limoges, B.S. llhorn, C. Degrand. Anal. Chem. 69 (1997) 4688.

[7] M. Albareda-Sirvent, A. Merkoci, S. Alegret, Sensors and Actuators B 69 (2000) 153.

[8] J. WANG, B. TIAN, E. SAHLIN. Anal Chem 71 (1999)3901.

[9] O. BAGEL, E. L'HOSTIS, G. LAGGER, MD. OSBORNE, BJ. SEDDON, HH. GIRAULT, J. Electroanal. Chem. 469 (1999) 189.

[10] R. McGLENNEN, Clin Chem 47 (2001) 393.

[11] JP. HART, SA. WRING. Trends Anal Chem 16 (1997) 89.

[12] M. ALBAREDA-SIRVENT, A. MERKcI, S. ALEGRET. Sens Actuators B 69 (2000) 153.

[13] J. Wang, *Analyst*, 119 (1994) 763.

[14] J. Wang and B.Tian, *Anal. Chem.* 64 (1992) 1706.

[15] J.Wang, B.Tian, *Anal. Chem.* 65 (1993) 1529.

[16] J.Wang, B.Tian, *Anal. Chim. Acta*, 274 (1993) 1.

[17] J.Wang, J.Lu, B.Tian, C.Yarnitzsky, J. Electroanal. Chem. 361 (1993) 77.

[18] J.P.Hart, S.A.Wring, Electroanalysis, 6 (1994) 617.

[19] K.Z. Brainina, Anal. Chim. Acta, 305 (1995) 146.

[20] Kh.Z. Brainina, H. Shaffer, A.V. Ivanona, N.A. Malakhova, Anal. Chim. Acta, 330 (1996) 175.

[21] D. Desmond, B. Lane, J. Alderman, M. Hill, D.W.M. Arrigan, J.D. Glennon, Sens. Actuators B, 48 (1998) 409.

[22] I. Palchetti, A. Cagnini, M. Mascini, A.P.F. Turner, Mikrochim. Acta 131 (1999) 65.

[23] J. Wang, Z. Taha, Electroanalysis, 2 (1990) 383.

[24] B. Hoyer, M.T. Florence, G.E. Batley, Anal Chem, 59, 13 (1987) 1608.

[25] B. Hoyer, T.M. Florence, Anal. Chem. 59, 24 (1987) 2839.

[26] C.M.A. Brett, D.A. Fungano, Talanta, 50 (2000) 1223.

[27] Kh.Z. Brainina, A.V. Ivanova, N.A. Malakhova, Anal. Chim. Acta, 349 (1997) 85.

[28] N.F. Zakharchuk, S.Yu. Saraeva, N.S. Borisova, Kh.Z. Brainina, Electroanalysis, 11, 9 (1999) 614.

[29] C. Faller, G. Henze, N. Stojko, S. Saraeva, K.Z. Brainina, Fresenius J. Anal. Chem. 358 (1997) 670.

[30] R.R. Dalagin, H. Gunasingham, Anal. Chim. Acta, 291 (1994) 81.

[31] C. Macca, M. Bradshaw, A. Merkoci, G. Scollary, Anal. Letters, 30 (1997) 1223.

[32] M.Wojciechowski, J. Balcerzak, Analytical Chemistry, 62 (1990) 1325.

[33] M.Wojciechowski, W.Go, J.Osterjoung, Analytical Chemistry, 57 (1985) 155.

[34] P.T. Kissinger, W.R.Heinemann, Laboratory techniques in electroanalytical chemistry, ed P.T.Kissinger Heinemann, W.R. New York: Marcel Dekker, Inc, (1984).

[35] A. Cagnini, I. Palchetti, M. Mascini, A.P.F. Turner, Mikrochim. Acta, 121 (1995) 155.

[36] MODERN POLAROGRAPHIC AND VOLTAMMETRIC ANALYSIS. A HANDBOOK. AMEL instruments. Version 3.00. July 1994.

GENERAL CONCLUSION

The work prësentës in this тётоие has led to the dëvelopment of innovative analytical mëthods, based in particular on the development of sensitive modifiëed electrodes for the analysis of heavy a ГёШ trace mëtals.

In the course of this work we were able to develop in chapter 3 an ampërométric method for the determination of traces of mercury using a rapid, simple to implement and inexpensive test. This determination is based on the influence of mercury (II) ions on the tyrosine oxidation current. A decrease in the oxidation current of amiiK' acid is due to the formation of a complex between tyrosine and mercury ions on the surface of a platinum ëlectrode.

A ëlectrochemical study of the formation of the complex a ëlë effective after optimisation of the various paramëtres. The data obtained allowed the dëduction of the strechiomëtrie and the calculation of the complex formation constant. Ami^ acid L-Tyrosine showed significant sëlectivity towards mercury ions compared with cystëine.

The modification of carbon paste ëlectrodes with a polymëre constituted the second part of this work. We mass-modified the carbon paste with the monomëre 1,8-DAN which we subsequently polymërisë within the paste in various acidic media.

The conductive polymer obtained within the carbon paste has been compared with the same polymer synthesised on the surface of solid electrodes and on the surface of carbon paste electrodes. The ëlectropolymërisation is carried out in an acid solution containing the monomëre. A radical change has ëlë observed between the two polymërisations, the polymëre obtained on the carbon or platinum Electrode surface is a non-conductive film whereas it is a conductive film within the carbon paste. After characterising the synthëtisë polymer within the carbon paste in an acid medium, we investigated the possibility of using it as an electrochemical sensor capable of preconcentrating Pb (II) ions in an open circuit using the DPASV technique. After optimising the various parameters, we have shown that this new modification method provides a high-performance analysis of lead ions.

The development of sensitive screen-printed sensors was the subject of the third part of this work. These electrodes, obtained by printing a carbon and silver-based ink onto a flexible polymer support, are attractive tools, since they can be easily mass-produced and have a wide range of geometries, while being modestly priced.

We have modified these electrodes with a mercury acetate salt that we have protected with a polymer. This modification, used as a good alternative to the mercury thin film, is essentially aimed at developing new electrochemical sensors that are less harmful and present no environmental problems.

These sensors, ready for use in the field, enable the simultaneous analysis of lead and cadmium and all the other metals that form an amalgam with mercury. Analysis of the performance of the different polymers studied showed that the Methocel - mercury salt membrane performed best, especially for use in very acidic environments.

Serigraphy electrodes modified with a Methocel - mercury salt membrane have enabled a detection limit of the order of ppb and have been successfully tested for the analysis of various real samples.

Finally, and based on the results we have obtained in the course of this work, we can envisage three possible areas of research for this work:

- Gënëralisation of the polymerisation process with the different diaminonaplitalenes which

can be polymerised within the carbon paste electrode. Their affinity to metals remains to be studied.

- The incorporation of 1.8 DAN into the graphite ink during the preparation of a crimped electrode could be studied to improve the performance of the technique.
- The design of portable devices coupled to screen-printed electrodes that can be used in the field. Especially with the recent commercialisation of increasingly sophisticated instrumentation and modern miniaturised analytical equipment.

MIX
Papier aus verantwortungsvollen Quellen
Paper from responsible sources
FSC® C105338

Printed by Books on Demand GmbH, Norderstedt / Germany